Algebra through practice

**Book 1: Sets, relations and mappings**

# Algebra through practice

*A collection of problems in algebra with solutions*

# Book 1
# Sets, relations and mappings

T.S.BLYTH ○ E.F.ROBERTSON

*University of St Andrews*

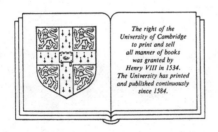

The right of the
University of Cambridge
to print and sell
all manner of books
was granted by
Henry VIII in 1534.
The University has printed
and published continuously
since 1584.

CAMBRIDGE UNIVERSITY PRESS
*Cambridge*
*London New York New Rochelle*
*Melbourne Sydney*

CAMBRIDGE UNIVERSITY PRESS
Cambridge, New York, Melbourne, Madrid, Cape Town, Singapore, São Paulo

Cambridge University Press
The Edinburgh Building, Cambridge CB2 8RU, UK

Published in the United States of America by Cambridge University Press, New York

www.cambridge.org
Information on this title: www.cambridge.org/9780521272858

First published 1984

*A catalogue record for this publication is available from the British Library*

*Library of Congress Catalogue Card Number: 83-24013*

ISBN 978-0-521-27285-8 paperback

Transferred to digital printing 2007

# Contents

# Preface

The aim of this series of problem-solvers is to provide a selection of worked examples in algebra designed to supplement undergraduate algebra courses. We have attempted, mainly with the average student in mind, to produce a varied selection of exercises while incorporating a few of a more challenging nature. Although complete solutions are included, it is intended that these should be consulted by readers only after they have attempted the questions. In this way, it is hoped that the student will gain confidence in his or her approach to the art of problem-solving which, after all, is what mathematics is all about.

The problems, although arranged in chapters, have not been 'graded' within each chapter so that, if readers cannot do problem $n$ this should not discourage them from attempting problem $n+1$. A great many of the ideas involved in these problems have been used in examination papers of one sort or another. Some test papers (without solutions) are included at the end of each book; these contain questions based on the topics covered.

TSB, EFR
St Andrews

# Background reference material

Courses on abstract algebra can be very different in style and content. Likewise, textbooks recommended for those courses vary enormously, not only in notation and exposition but also in their level of sophistication. Here is a list of some major texts that are widely used and to which the reader may refer for background material. The subject matter of these texts covers all six books in the *Algebra through practice* series, and in some cases a great deal more. For the convenience of the reader there is given below an indication of which parts of which of these texts is most relevant to the appropriate chapters of this book.

[1] I. T. Adamson, *Introduction to Field Theory*, Cambridge University Press, 1982.
[2] F. Ayres, Jr, *Modern Algebra*, Schaum's Outline Series, McGraw-Hill, 1965.
[3] D. Burton, *A First Course in Rings and Ideals*, Addison-Wesley, 1970.
[4] P. M. Cohn, *Algebra* Vol. I, Wiley, 1982.
[5] D. T. Finkbeiner II, *Introduction to Matrices and Linear Transformations*, Freeman, 1978.
[6] R. Godement, *Algebra*, Kershaw, 1983.
[7] J. A. Green, *Sets and Groups*, Routledge and Kegan Paul, 1965.
[8] I. N. Herstein, *Topics in Algebra*, Wiley, 1977.
[9] K. Hoffman and R. Kunze, *Linear Algebra*, Prentice Hall, 1971.
[10] S. Lang, *Introduction to Linear Algebra*, Addison-Wesley, 1970.
[11] S. Lipschutz, *Linear Algebra*, Schaum's Outline Series, McGraw-Hill, 1974.
[12] I. D. Macdonald, *The Theory of Groups*, Oxford University Press, 1968.
[13] S. MacLane and G. Birkhoff, *Algebra*, Macmillan, 1968.
[14] N. H. McCoy, *Introduction to Modern Algebra*, Allyn and Bacon, 1975.
[15] J. J. Rotman, *The Theory of Groups: An Introduction*, Allyn and Bacon, 1973.
[16] I. Stewart, *Galois Theory*, Chapman and Hall, 1973.

[17] I. Stewart and D. Tall, *The Foundations of Mathematics*, Oxford University Press, 1977.

## References useful to Book 1

1: Sets    [2, Chapter 1], [6, Chapters 1, 3], [7, Chapter 1], [17, Chapter 3].
2: Relations    [2, Chapter 2], [6, Chapter 4], [7, Chapter 2], [17, Chapter 4].
3: Mappings    [6, Chapter 2], [7, Chapter 3], [17, Chapter 5].

In [2] the author writes mappings on the right, and uses 'one-to-one' for injective and 'onto' for surjective. In [6] the author uses ⊂ for set inclusion (where we use ⊆). In [7] mappings are written on the right. In [17] the definition of (partial) order differs from ours in that the axiom of reflexivity is missing.

# 1: Sets

We assume that the reader has a basic knowledge of elementary set theory and we shall use standard (i.e. the most commonly accepted) notation. Thus, for example, we shall denote the complement of a subset $A$ of a set $E$ simply by $A'$ except when confusion can occur in which case we shall write $C_E(A)$. If $A$ and $B$ are subsets of $E$ then the difference set $A \cap B'$ will be denoted by $A \setminus B$ (some authors use $A - B$), and the symmetric difference set $(A \cap B') \cup (A' \cap B)$ will be denoted by $A \vartriangle B$.

Some questions in this section are best dealt with using the algebra of set theory, with which we assume that the reader is familiar. For example, this includes the distributive laws

$$A \cap (B \cup C) = (A \cap B) \cup (A \cap C)$$

and

$$A \cup (B \cap C) = (A \cup B) \cap (A \cup C),$$

and the de Morgan laws

$$(A \cap B)' = A' \cup B' \quad \text{and} \quad (A \cup B)' = A' \cap B'.$$

Other questions, particularly those dealing with set-theoretic identities, are best dealt with using Venn diagrams.

Other standard notation that we shall employ includes $\mathbf{P}(E)$ for the power set of $E$ (i.e. the set of all subsets of $E$); $|A|$ for the number of elements in the set $A$; $A \times B$ for the cartesian product of $A$ and $B$ (i.e. the set of ordered pairs $(a, b)$ with $a \in A$ and $b \in B$); and the following for particular subsets of the number system:

$\text{IN} = \{0, 1, 2, \ldots\}$ for the set of natural numbers;

$\mathbb{Z} = \{\ldots, -2, -1, 0, 1, 2, \ldots\}$ for the set of integers;

$\mathbb{Q} = \{a/b \mid a, b \in \mathbb{Z}, b \neq 0\}$ for the set of rationals;

1

IR for the set of real numbers;

$\mathbb{C}$ for the set of complex numbers;

$]a, b] = \{x \in IR \mid a < x \leqslant b\}$;

$]a, b[ = \{x \in IR \mid a < x < b\}$, etc.

Finally the usual logical abbreviations $\exists$ (there exists), $\forall$ (for all), $\Rightarrow$ (implies), $\Leftrightarrow$ (if and only if) will be used throughout, both in the problems and in their solutions.

1.1   Let $A = \{\emptyset, \{\emptyset\}, \{\emptyset, \{\emptyset\}\}\}$. Which of the following are true?

$\emptyset \subseteq A$,   $\emptyset \in A$,   $\{\emptyset\} \in A$,   $\{\emptyset\} \subseteq A$,   $\{\{\emptyset\}\} \subseteq A$,   $\{\{\emptyset\}, \emptyset\} \subseteq A$,

$\{\{\emptyset\}, \emptyset\} \in A$.

1.2   List the elements of $P(P(\emptyset))$ and of $P(P(P(\emptyset)))$.

1.3   For the set $E = \{1, \{1\}, 2, \{1, 2\}\}$ determine $P(E)$ and $E \cap P(E)$.

1.4   Find four examples of a set $A$ with the property that every element of $A$ is a subset of $A$.

1.5   Can you find sets $A, B, C$ such that

$A \subseteq B \in C$   and   $A \in B \subseteq C$?

1.6   Which of the following hold for all sets $A, B$ and $C$?
   (a)  If $A \notin B$ and $B \notin C$ then $A \notin C$.
   (b)  If $A \neq B$ and $B \neq C$ then $A \neq C$.
   (c)  If $A \in B$ and $B \not\subseteq C$ then $A \notin C$.
   (d)  If $A \subseteq B$ and $B \subseteq C$ then $C \not\subseteq A$.
   (e)  If $A \subseteq B$ and $B \in C$ then $A \notin C$.
   (f)  If $A \cap C \subseteq B$ then $(A \cap B) \cup (B \cap C) = B$.
   (g)  If $A \cap C \in B$ then $A \in B \cup C$.
   (h)  If $A \cap C = \emptyset$ and $B \cap C = \emptyset$ then $(A \cup B) \cap C = \emptyset$.

1.7   Show that $\{x, y\} \cap \{y, z\} = \{y\}$ may be false.

1.8   Let $A, B, C$ be subsets of a set $X$. Simplify the expressions
   (a)  $(A \cup (A \cup B)')'$;
   (b)  $((A \cup \emptyset) \cap (B \cup A') \cap (A \cup B' \cup X))'$;
   (c)  $(A \cup (B \cap C) \cup (B' \cap C') \cup C)'$.

1.9   Let $A, B$ be subsets of a set $X$. Prove, using a Venn diagram, that

$(A \cap B') \cup (A' \cap B) = A \cup B \Leftrightarrow A \cap B = \emptyset$.

2

**1.10**  Let $A, B, C$ be subsets of a set $X$. Prove, using Venn diagrams, that
      (*a*) $A \triangle (B \triangle C) = (A \triangle B) \triangle C$;
      (*b*) $A \cup B = A \triangle B \triangle (A \cap B)$;
      (*c*) $A \cap (B \triangle C) = (A \cap B) \triangle (A \cap C)$;
      (*d*) $A \triangle (B \cap C) = (A \triangle B) \cap (A \triangle C)$ if and only if $A \cap B = A \cap C$.

**1.11**  If $A, B$ are subsets of a set $E$ let $A \mid B = C_E(A \cap B)$. Show that $C_E(A) = A \mid A$ and that $A \cap B = (A \mid B) \mid (A \mid B)$. Express $\cup$ in terms of $\mid$ alone.

**1.12**  Let $A, B, C, D$ be sets with $\{A, B\} = \{C, D\}$. Prove that $A \cap B = C \cap D$ and that $A \cup B = C \cup D$.

**1.13**  If $A, B, C$ are sets prove that $A \cap (B \cup C) \subseteq (A \cap B) \cup C$, with equality if and only if $C \subseteq A$.

**1.14**  For sets $E, F, G$ prove, using a Venn diagram, that
$$(E \cap F) \cup (F \cap G) \cup (G \cap E) = (E \cup F) \cap (F \cup G) \cap (G \cup E).$$

**1.15**  Give an example of sets $A, B, C, D$ with
$$(A \cup B) \times (C \cup D) \neq (A \times C) \cup (B \times D).$$

**1.16**  Given two objects $x, y$ one may define the ordered pair $(x, y)$ by
$$(x, y) = \{\{x\}, \{x, y\}\}.$$
Use this definition to prove that $(x, y) = (x^*, y^*)$ if and only if $x = x^*$ and $y = y^*$. Prove also that, for every object $x$,
$$\{x\} \times \{x\} = \{\{\{x\}\}\}.$$

**1.17**  Prove that if $A, B, C$ are sets with $A$ and $B$ not empty then
$$(A \times B) \cup (B \times A) = C \times C \Leftrightarrow A = B = C.$$

**1.18**  Let $\mathscr{F}$ be a collection of sets such that
$$X, Y \in \mathscr{F} \Rightarrow X \setminus Y \in \mathscr{F}.$$
Prove that if $X, Y \in \mathscr{F}$ then $X \cap Y \in \mathscr{F}$.

**1.19**  Let $\mathscr{F}$ be a collection of sets. Define
$$\mathscr{F}^0 = \{X \setminus Y \mid X, Y \in \mathscr{F}\}.$$
Show that $\mathscr{F}^0 \subseteq (\mathscr{F}^0)^0$. Give an example to show that it is possible to have $\mathscr{F}^0 \neq (\mathscr{F}^0)^0$.

**1.20**  Let $A, B$ be sets. Are the following true?
      (*a*) $P(A) \cap P(B) = P(A \cap B)$;
      (*b*) $P(A) \cup P(B) = P(A \cup B)$.

3

**1.21**  If $A, B, C$ are sets prove, using Venn diagrams, that $A \setminus (B \cup C) = (A \setminus B) \cup (A \setminus C)$ if and only if $A \triangle (B \cup C) = (A \triangle B) \cup (A \triangle C)$. Find sets $A, B, C$ with

$$A \triangle (B \cup C) = (A \triangle B) \cup (A \triangle C);$$
$$A \triangle (B \cup C) \neq (A \triangle B) \cup (A \triangle C).$$

**1.22**  Let $A_1, \ldots, A_m$ be subsets of a set $E$. Define $B_1, \ldots, B_m$ recursively as follows:

$$B_1 = A_1, \quad (\forall n \geqslant 2)\, B_n = A_n \setminus \bigcup_{k=1}^{n-1} A_k.$$

Show that $B_i \cap B_j = \emptyset$ for $i \neq j$ and that $\bigcup_{i=1}^{m} B_i = \bigcup_{i=1}^{m} A_i$.

**1.23**  Let $A, B$ be sets with $A \subseteq B$. Prove that there is a unique subset $X$ of $B$ such that $X \cup A = B$ and $X \cap A = \emptyset$.

**1.24**  An examination in three subjects algebra $(A)$, biology $(B)$, chemistry $(C)$ was taken by 41 students. The following table shows how many students failed the various combinations of subjects.

| Subjects | $A$ | $B$ | $C$ | $A, B$ | $A, C$ | $B, C$ | $A, B, C$ |
|---|---|---|---|---|---|---|---|
| No. of failed students | 12 | 5 | 8 | 2 | 6 | 3 | 1 |

How many students passed all three subjects?

**1.25**  At least 70% of a class of students study algebra, at least 75% study calculus, at least 80% study geometry, and at least 85% study trigonometry. What percentage (at least) must study all four subjects?

**1.26**  Let $E$ be a set consisting of $n$ elements. If $X, Y$ are subsets of $E$ such that the number of elements in the sets $X \cap Y, X' \cap Y, X \cap Y', X' \cap Y'$ are $p, q, r, s$ respectively, prove that $p + q + r + s = n$.

   In a sixth form of $n$ girls and boys each pupil is either an arts student or a science student. If the proportion of arts students among the girls is greater than the proportion of arts students among the boys, show that the proportion of girls among the arts students is greater than the proportion of girls among the science students.

**1.27**  If $A = \{x \in \mathbb{Z} \mid (\exists y \in \mathbb{Z}) x = 2y\}$ and $B = \{a \in \mathbb{Z} \mid (\exists b, c \in \mathbb{Z}) a = 6b + 10c\}$ prove that $A = B$.

**1.28**  For which $S \in \{\mathbb{N}, \mathbb{Z}, \mathbb{Q}, \mathbb{R}, \mathbb{C}\}$ are the following statements true?
   (a) $\{x \in S \mid x^2 = 5\} \neq \emptyset$;

(b) $\{x \in S \mid |x - 1| \leqslant \frac{1}{2}\} = \{1\}$;

(c) $\{x \in S \mid x^2 = -1\} = \emptyset$.

**1.29**  Given $n \in \mathbb{N}$ define $n\mathbb{Z} = \{nx \mid x \in \mathbb{Z}\}$. Is it true that given $n_1, n_2 \in \mathbb{N}$ there exists $m \in \mathbb{N}$ with

$$n_1 \mathbb{Z} \cap n_2 \mathbb{Z} = m\mathbb{Z}?$$

Is it true that, for some $m \in \mathbb{N}$,

$$n_1 \mathbb{Z} \cup n_2 \mathbb{Z} = m\mathbb{Z}?$$

**1.30**  Let $A = \{x \in \mathbb{N} \mid 1 \leqslant x \leqslant n\}$.

      (a)  How many subsets does $A$ have?

      (b)  How many subsets of $A$ contain at least one even integer?

      (c)  How many subsets of $A$ contain exactly one even integer?

(*Hint:* consider separately the cases $n$ even, $n$ odd.)

**1.31**  Let $A = \{x \in \mathbb{N} \mid 1 \leqslant x \leqslant n\}$. What is the maximum possible $k$ for which $A_i \subseteq A$ $(i = 1, \ldots, k)$ and $A_i \subset A_j$ if $i < j$? Find $\sum_{i=1}^{k} |A_i|$.

**1.32**  Express as a union of intervals

$$\left\{ x \in \mathbb{R} \setminus \{-1, 4\} \;\middle|\; \frac{1}{(x+1)(x-4)} > -\frac{1}{4} \right\}.$$

**1.33**  Express as a union of intervals the set of real numbers $k$ for which

$$\left\{ x \in \mathbb{R} \;\middle|\; \frac{(x-1)^2}{(x+1)(x+3)} = k \right\} = \emptyset.$$

# 2: Relations

A relation $R$ between a set $E$ and a set $F$ is a subset of $E \times F$, and we shall use the notation $(x, y) \in R$ or $xRy$ with the same meaning. When we think of a coordinate pictorial representation of $E \times F$ we refer to $\{(x, y) \mid xRy\}$ as the graph of $R$, to $\{x \in E \mid (\exists y \in F)(x, y) \in R\}$ as the domain of $R$, and to $\{y \in F \mid (\exists x \in E)(x, y) \in R\}$ as the image of $R$. A relation between $E$ and $E$ is called a (binary) relation on $E$.

If the relation $R$ on $E$ is reflexive ($xRx$ for all $x \in E$), symmetric (if $xRy$ then $yRx$), and transitive (if $xRy$ and $yRz$ then $xRz$), then $R$ is an equivalence relation on $E$. When $R$ is an equivalence relation on $E$ we sometimes write $x \equiv y(R)$ instead of $xRy$. For $x \in E$ the $R$-class of $x$, i.e. $\{y \in E \mid yRx\}$, is denoted by $[x]_R$ or simply $[x]$ when no confusion can arise. The following are equivalent:

$$x \equiv y(R), \quad y \in [x]_R, \quad [x]_R = [y]_R, \quad [x]_R \cap [y]_R \neq \emptyset.$$

It follows that two $R$-classes either are disjoint (i.e. have empty intersection) or are identical. This leads to the notion of a partition of $E$ as a collection of non-empty subsets of $E$ which are pairwise disjoint and whose union is the whole of $E$. If $R$ is an equivalence relation on $E$ then the $R$-classes form a partition of $E$. Conversely, every partition of $E$ defines an equivalence relation $\equiv$ on $E$ by

$$x \equiv y \Leftrightarrow x, y \text{ belong to the same subset in the partition.}$$

An example of an equivalence relation is the relation mod $n$ defined on $\mathbb{Z}$ by $a \equiv b(\bmod n)$ if and only if $n$ divides $a - b$. The corresponding partition consists of the equivalence classes

## 2: Relations

$$[0] \quad = \{\ldots, -2n, -n, 0, n, 2n, \ldots\}$$
$$[1] \quad = \{\ldots, -n+1, 1, n+1, \ldots\}$$
$$\vdots$$
$$[n-1] = \{\ldots, -1, n-1, 2n-1, \ldots\}$$

A relation $R$ on a set $E$ that is reflexive, anti-symmetric (if $xRy$ and $yRx$ then $x = y$), and transitive is called an order (or a partial order) and is often written $\leqslant$. When $a \leqslant b$ and $a \neq b$ we write $a < b$. Examples of order relations are

    (a) $\subseteq$ on $\mathbf{P}(E)$;
    (b) $\mid$ on $\mathbb{Z}$, where $a \mid b \Leftrightarrow a$ divides $b$.

An order relation can often be represented pictorially by a Hasse diagram. In this, $a < b$ is exhibited by joining the point representing $a$ to the point representing $b$ by an increasing line segment. For example, the Hasse diagrams for the set $\{1, 2, 3, 4\}$ ordered first in the usual way and then by divisibility are shown in Figs 2.1 (a) and (b) respectively.

**Fig.2.1**

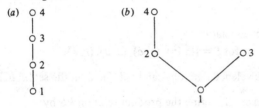

**2.1** Let $S$ be the relation defined on $\mathbb{R}$ by
$$xSy \Leftrightarrow x^2 = x|y+1|.$$
Sketch the graph of $S$.

**2.2** Sketch the graphs of the following relations on $\mathbb{R}$:
    (a) $\{(x, y) \mid |x+y| \leqslant 1\}$;
    (b) $\{(x, y) \mid 2x^2 + 3xy - 2y^2 \leqslant 0\}$;
    (c) $\{(x, y) \mid (x-y)(x-2y)(x-3y) \geqslant 0\}$;
    (d) $\{(x, y) \mid x+y-4 \leqslant 0, 2x-y-4 \leqslant 0, 2x-5y-10 \leqslant 0,$
       $3x-y+3 \geqslant 0\}$.

**2.3** Let $A = \{1, 2, 3, 4\}$. Determine the graphs of the relations $R$, $S$ defined

7

on $A$ by

$$aRb \Leftrightarrow a + b \leqslant 4;$$
$$aSb \Leftrightarrow a(b+1) \leqslant 6.$$

**2.4** Relations $R_1$ and $R_2$ are defined on $\mathbb{R}$ by

$$xR_1y \Leftrightarrow -10 \leqslant x + 5y \leqslant 10,$$
$$xR_2y \Leftrightarrow x^2 + y^2 \leqslant 4, x \geqslant y.$$

Sketch the graphs of these relations.

**2.5** Let the relation $\rho$ on a set $A$ have the properties

    (*a*) $a\rho a$ for every $a \in A$;

    (*b*) if $a\rho b$ and $b\rho c$ then $c\rho a$.

Prove that $\rho$ is an equivalence relation on $A$. Does every equivalence relation on $A$ satisfy (*a*) and (*b*)?

**2.6** Consider the relation $R = \{(a, b), (a, c), (a, a), (b, d), (c, c)\}$ defined on the set $X = \{a, b, c, d\}$. Find the minimum number of elements of $X \times X$ which must be adjoined to $R$ in order to make $R$

    (*a*) reflexive;

    (*b*) symmetric;

    (*c*) an equivalence relation.

Answer the same questions for $S = \{(a, b), (a, c), (a, a), (c, c)\}$.

**2.7** How many different equivalence relations can be defined on the set $\{a, b, c\}$?

**2.8** Given relations $R, S$ on a set $A$, define the product relation $RS$ by

$$(x, y) \in RS \Leftrightarrow (\exists z \in A)((x, z) \in S \text{ and } (z, y) \in R).$$

Give an example of relations $R, S$ with $RS = SR$ and an example of relations $R, S$ with $RS \neq SR$.

    Prove that if $R$ and $S$ are equivalence relations then $RS$ is an equivalence relation if and only if $RS = SR$. Deduce that $RS$ is an equivalence relation if and only if $SR$ is an equivalence relation.

**2.9** Let $R_1, R_2$ and $S$ be relations on a set $X$. Prove that

    (*a*) if $R_1 \subseteq R_2$ then $SR_1 \subseteq SR_2$ and $R_1S \subseteq R_2S$;

    (*b*) $S(R_1 \cup R_2) = SR_1 \cup SR_2$.

**2.10** If $R_1$ and $R_2$ are equivalence relations on a set $X$ prove that

    (*a*) $R_1 \cap R_2$ is an equivalence relation;

    (*b*) $R_1 \cup R_2$ need not be an equivalence relation.

Give an example of equivalence relations $R_1$ and $R_2$ with $R_1 \neq R_2$ and $R_1 \cup R_2$ an equivalence relation.

**2.11** Let $\alpha$ be the relation on $\mathbb{N}$ defined by

$$a\alpha b \Leftrightarrow a^2 \equiv b^2 (\text{mod } 7).$$

Show that $\alpha$ is an equivalence relation. Into how many equivalence classes does $\alpha$ partition $\mathbb{N}$?

**2.12** Let $S = \mathbb{R} \setminus \{0\}$. Define a relation $\rho$ on $S \times S$ by

$$(a, b)\rho(c, d) \Leftrightarrow c^2 b = a^2 d.$$

Prove that $\rho$ is an equivalence relation. Describe geometrically the $\rho$-classes.
If $\sigma$ is defined on $S \times S$ by

$$(a, b)\sigma(c, d) \Leftrightarrow c^4 b^2 = a^4 d^2,$$

show that $\sigma$ is an equivalence relation. Describe geometrically the $\sigma$-classes. Explain how the equivalence classes of $\rho$ and of $\sigma$ are related.
If the relation $\tau$ is defined on $\mathbb{R} \times \mathbb{R}$ by

$$(a, b)\tau(c, d) \Leftrightarrow c^2 b = a^2 d,$$

is $\tau$ an equivalence relation?

**2.13** Consider the relation $\sim$ defined on $\mathbb{C} \setminus \{0\}$ by

$$z_1 \sim z_2 \Leftrightarrow |z_1|(|z_2|^2 + 1) = |z_2|(|z_1|^2 + 1).$$

Prove that $\sim$ is an equivalence relation. If $a \in \mathbb{R}$ is such that $0 < a < 1$, sketch on the Argand diagram the $\sim$-class of $a$.

**2.14** Consider the relation $\sim$ defined on $\mathbb{C} \setminus \{0\}$ by

$$z_1 \sim z_2 \Leftrightarrow z_1 \bar{z}_1 (z_2 + \bar{z}_2) = z_2 \bar{z}_2 (z_1 + \bar{z}_1).$$

Prove that $\sim$ is an equivalence relation. If $a$ is a non-zero number on the real axis, give a geometrical description of the $\sim$-class of $a$.

**2.15** Let $S = \{(x, y) \in \mathbb{R} \times \mathbb{R} \mid x \neq 0, y \neq 0\}$ and define a relation $\sim$ on $S$ by

$$(x_1, y_1) \sim (x_2, y_2) \Leftrightarrow x_1 y_1 (x_2^2 - y_2^2) = x_2 y_2 (x_1^2 - y_1^2).$$

(a) Show that $\sim$ is an equivalence relation.
(b) If $(a, b)$ is a fixed element of $S$ show that

$$(x, y) \sim (a, b) \Leftrightarrow \frac{y}{x} = \frac{b}{a} \quad \text{or} \quad \frac{y}{x} = -\frac{a}{b}.$$

(c) Sketch the $\sim$-class containing $(2, 1)$.

**2.16** If $a$ is a given real number consider the relation $f$ on $\mathbb{R}$ given by

$$(x, y) \in f \Leftrightarrow y = x^2 + ax + a^2.$$

(*a*) By considering the graph of $f$ prove that if $A = \{x \in \mathbb{R} \mid (x, 1) \in f\}$ then

        (i) $A = \emptyset$ if and only if $|a| > 2/\sqrt{3}$;

        (ii) $|A| = 1$ if and only if $|a| = 2/\sqrt{3}$;

        (iii) $|A| = 2$ if and only if $|a| < 2/\sqrt{3}$.

(*b*) Prove that the relation $S$ defined on $\mathbb{R}$ by

$$x \equiv y(S) \Leftrightarrow x^3 - y^3 = x - y$$

is an equivalence relation. Deduce from the above that the $S$-class of $x \in \mathbb{R}$ consists of

        (i) a single element if and only if $|x| > 2/\sqrt{3}$;

        (ii) two elements if and only if $|x| = 2/\sqrt{3}$ or $|x| = 1/\sqrt{3}$;

        (iii) three elements otherwise.

**2.17**    (*a*) Prove that the relation $R$ defined on $\mathbb{R}$ by

$$xRy \Leftrightarrow x^2 - y^2 = 2(y - x)$$

is an equivalence relation. Determine the $R$-class of 0 and the $R$-class of 1.

(*b*) The following argument leads to a false conclusion. Explain where it is incorrect.

Since $x^2 - y^2 = (x + y)(x - y)$ it follows that if $x^2 - y^2 = 2(y - x)$ then $(x + y)(x - y) = -2(x - y)$ and so $x + y = -2$. Hence the relation $S$ defined by $xSy \Leftrightarrow x + y = -2$ is also an equivalence relation, and from 1$S$1 we have $2 = -2$.

**2.18**    Let $M$ be a set of $mn + 1$ positive integers. Let $\equiv$ be the relation on $M$ defined by

$$a \equiv b \Leftrightarrow a \mid b,$$

and let $S$ be the relation on $M$ defined by

$$aSb \Leftrightarrow a \not\equiv b \quad \text{and} \quad b \not\equiv a.$$

Show that $M$ contains either a subset $\{a_1, a_2, \ldots, a_{m+1}\}$ with $a_i \equiv a_{i+1}$ for $1 \leqslant i \leqslant m$ or a subset $\{b_1, b_2, \ldots, b_{n+1}\}$ with $b_j S b_k$ for $j \neq k$. (*Note:* this question is quite hard.)

**2.19**    Let $A_1, A_2, \ldots, A_n$ be subsets of a set $X$. For each $A_i$, let $A_i^0$ denote $A_i$ and let $A_i^1$ denote the complement of $A_i$ in $X$. A *constituent* of $X$ with respect to $A_1, \ldots, A_n$ is defined to be a non-empty subset of the form

$$A_1^{\epsilon_1} \cap A_2^{\epsilon_2} \cap \ldots \cap A_n^{\epsilon_n}$$

where each $\epsilon_i$ is either 0 or 1. Prove that the constituents of $X$ are disjoint and that they partition $X$.

Suppose now that $A, B, C$ are subsets of $X$. Write the subset $A \setminus (B \setminus C)$ as a union of constituents of $X$ with respect to $A, B, C$. If $\equiv$ is the equivalence relation defined on $X$ by the partition of constituents, is it possible to have $x \equiv y$ when $x \in A \setminus (B \setminus C)$ and $y \in (A \cap B) \setminus C$?

**2.20** Define a relation $\rho$ on $\mathbb{R}_+ = \{x \in \mathbb{R} \mid x \geqslant 0\}$ by

$$(a, b) \in \rho \Leftrightarrow a - \sqrt{(a+1)} \leqslant b - \tfrac{1}{4} \leqslant a + \sqrt{(a+1)}.$$

Is $\rho$ reflexive? Is $\rho$ symmetric? Is $\rho$ transitive?

**2.21** A set contains 1000 elements and is partitioned into $m + 1$ subsets. The smallest subset in the partition contains $n$ elements, the largest contains $n + m$ elements, and no two members of the partition contain the same number of elements. Find all possible positive values of $m$ and $n$.

**2.22** Let $E = \{(x, y) \in \mathbb{R} \times \mathbb{R} \mid x \neq 0, y \neq 0\}$. Define a relation $\sim$ on $E$ by

$$(x, y) \sim (a, b) \Leftrightarrow (xb)^2 = (ya)^2.$$

Verify that $\sim$ is an equivalence relation. Prove that $(x, y) \sim (a, b)$ if and only if there is a non-zero real number $k$ such that $x = ka, y = \pm kb$. Sketch the $\sim$-class of the element $(2, 1)$.

**2.23** Let $E = \{(x, y) \in \mathbb{R} \times \mathbb{R} \mid x \neq 0, y \neq 0\}$. Define a relation $\equiv$ on $E$ by

$$(x, y) \equiv (z, t) \Leftrightarrow xy(z^2 + zt + t^2) = zt(x^2 + xy + y^2).$$

Verify that $\equiv$ is an equivalence relation on $E$. If $m \neq 0$ find the equivalence class of $(1, m)$. Hence describe geometrically all the equivalence classes.

**2.24** Let $a, b \in \mathbb{Z}$ and suppose that $a, b$ are coprime with $a > b > 0$. Define recursively $f_1, f_2, \ldots$ by $f_1 = a, f_2 = a - b$ and, for $k \geqslant 2$,

$$f_{k+1} = \begin{cases} f_k + a & \text{if} \quad f_k < b; \\ f_k - b & \text{if} \quad f_k \geqslant b. \end{cases}$$

Prove that $\{f_1, f_2, \ldots, f_{a+b}\}$ forms a 'complete set of representatives' mod $a + b$, in the sense that each $(\bmod\, a + b)$-class contains one and only one $f_i$.

What happens if $a$ and $b$ fail to be coprime?

**2.25** Draw the Hasse diagrams for the set $E = \{1, 2, \ldots, 10\}$ when ordered by
   (a) divisibility;
   (b) the relation $\leqslant$ defined by $x \qquad y$ if and only if either $x = y$ or $x \prec y$ where, for all positive integers $p$ and $q$,

$$2p + 1 \prec 2q;$$
$$2p + 1 \prec 2q + 1 \Leftrightarrow p < q;$$
$$2p < 2q \Leftrightarrow q < p.$$

(c) the relation $\leqslant$ defined by $p \leqslant q$ if and only if $p = q$ or $p < q$ where

$$p < q \Leftrightarrow \begin{cases} \text{there is a prime } t \in E \text{ such} \\ \text{that } p = t^a, q = t^b, a < b. \end{cases}$$

**2.26** Draw the Hasse diagrams for $\mathbf{P}(E)$ ordered by $\subseteq$ when

(a) $E = \emptyset$;

(b) $E = \{\emptyset\}$;

(c) $E = \{\{\emptyset\}\}$;

(d) $E = \{\emptyset, \{\emptyset\}\}$;

(e) $E = \{\emptyset, \{\emptyset\}, \{\emptyset, \{\emptyset\}\}\}$.

**2.27** Draw the Hasse diagram for the set of positive divisors of 210 ordered by divisibility.

**2.28** Let $E = \{x_{ij} \mid 1 \leqslant i \leqslant m, \ 1 \leqslant j \leqslant n\}$ be a set of $mn$ distinct positive real numbers. For every $i \in [1, m]$ define $y_i = \max \{x_{ij} \mid 1 \leqslant j \leqslant n\}$, and for every $j \in [1, n]$ define $z_j = \min \{x_{ij} \mid 1 \leqslant i \leqslant m\}$. Prove that

$$\min \{y_i \mid 1 \leqslant i \leqslant m\} \geqslant \max \{z_j \mid 1 \leqslant j \leqslant n\}.$$

A regiment of soldiers, each of a different height, stands at attention in a rectangular array. Of the soldiers who are the tallest in their row, the smallest is Sergeant Mintall; and of the soldiers who are the smallest in their column, the tallest is Corporal Max Small. How does the height of Sergeant Mintall compare with that of Corporal Max Small?

**2.29** Let $E$ be a set on which there is defined an order relation $\leqslant$. Let $\equiv$ be an equivalence relation on $E$. Suppose that the conditions $x \equiv z$ and $x \leqslant y \leqslant z$ together imply that $x \equiv y \equiv z$ for $x, y, z \in E$. Let $\bar{E} = \{[x] \mid x \in E\}$ and define a relation $R$ on $\bar{E}$ by

$$R = \{([x], [y]) \mid (\forall a \in [x])(\exists b \in [y]) a \leqslant b\}.$$

Prove that $R$ is an order relation on $\bar{E}$.

# 3: Mappings

A mapping (or function) $f$ from a set $X$ to a set $Y$ is a relation between $X$ and $Y$ (i.e. a subset of $X \times Y$) with the properties
    (a)  given $x \in X$ there is some $y \in Y$ with $(x, y) \in f$,
    (b)  if $(x, y_1), (x, y_2) \in f$ then $y_1 = y_2$.
We denote such a mapping $f$ by writing $f : X \to Y$, and if $(x, y) \in f$ we use the notation $y = f(x)$. The set $X$ is called the domain of $f$; and the subset $\{y \in Y \mid (\exists x \in X) y = f(x)\}$ of $Y$ is called the image of $f$ and is denoted by Im $f$. When it is clear what the sets $X$ and $Y$ are, we often write the mapping $f$ in the form $x \to f(x)$. If $A \subseteq X$ then $f(A)$ is defined to be $\{y \in Y \mid (\exists x \in A) y = f(x)\}$. In particular, $f(X) = $ Im $f$. Two mappings $f : A \to B$ and $g : C \to D$ are said to be equal if $A = C, B = D$ and $(\forall x \in A) f(x) = g(x)$.

Given $f : X \to Y$ and $g : Y \to Z$, the composite $g \circ f$ is the mapping $g \circ f : X \to Z$ defined by $(g \circ f)(x) = g[f(x)]$ for every $x \in X$. For mappings $f, g$ and $h$ we recall that the associative law $(f \circ g) \circ h = f \circ (g \circ h)$ holds whenever these multiple composites are defined. Also, when $f : X \to X$ we use the usual notation $f^2$ for $f \circ f$, and in general $f^n$ for $f \circ f \circ \dots \circ f$ ($n$ terms).

A mapping $f : X \to Y$ is said to be bijective if it is injective (if $f(x_1) = f(x_2)$ then $x_1 = x_2$) and surjective (Im $f = Y$). In this case there is a unique mapping $f^{-1} : Y \to X$ such that $f^{-1} \circ f = \mathrm{id}_X$ and $f \circ f^{-1} = \mathrm{id}_Y$ where the notation $\mathrm{id}_X$ denotes the identity mapping on $X$, namely the mapping $\mathrm{id}_X : X \to X$ given by $\mathrm{id}_X(x) = x$ for all $x \in X$. The composite $f \circ g$ of two bijections $f$ and $g$ is also a bijection; we have $(f \circ g)^{-1} = g^{-1} \circ f^{-1}$. Some authors use the terms one-one to mean injective, and onto to mean surjective.

A relation on $\mathbb{R}$ is a mapping if and only if every line parallel to the $y$-axis meets the graph of the relation precisely once. Other geometrical criteria in terms of graphs can also be useful. For example, a mapping $f : \mathbb{R} \to \mathbb{R}$ is injective if and only if every line parallel to the $x$-axis meets the graph of $f$ in at most one point; and $f$ is surjective if and only if every line parallel to the $x$-axis meets the graph of $f$ in at least one point. Hence $f$ is bijective if and only if every line parallel to the $x$-axis meets the graph of $f$ precisely once.

By way of example, consider the mapping $f : \mathbb{R} \to \mathbb{R}$ given by $x \to f(x) = [\![x]\!]$ where $[\![x]\!]$ is defined to be the greatest integer that is less than or equal to $x$. The graph of $f$ is as shown in Fig. 3.1 (in which each arrow indicates a constant value throughout an interval of the form $[n, n+1[$).

**Fig.3.1**

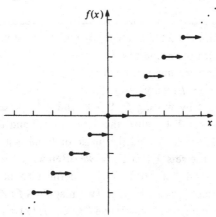

This mapping is not injective (for example, the line $y = 1$ meets the graph in infinitely many points), nor is it surjective (for example, the line $y = \frac{1}{2}$ does not meet the graph at all). Note, however, that $\operatorname{Im} f = \mathbb{Z}$ and that $f$ induces a surjective mapping from $\mathbb{R}$ to $\mathbb{Z}$ also given by the prescription $x \to [\![x]\!]$.

3.1   Let $S, P, C$ be the functions from $\mathbb{R}$ to $\mathbb{R}$ given by $S(x) = x^2$, $P(x) = 2^x$ and $C(x) = \cos x$. Express each of the following functions $f : \mathbb{R} \to \mathbb{R}$ in terms of $S, P, C$ using $\cdot$ and $\circ$ (where $f \cdot g$ is defined by $(f \cdot g)(x) = f(x)g(x)$):

(a) $f(x) = 2^{\cos x}$;     (f) $f(x) = \cos 2^{x + \cos x}$;

(b) $f(x) = \cos 2^x$;     (g) $f(x) = \cos (\cos x)^2$;

(c) $f(x) = (\cos x)^2$;     (h) $f(x) = (\cos 2^{\cos x})^2$;

(d) $f(x) = \cos x^2$;     (i) $f(x) = 2^{2 \cos x}$;

(e) $f(x) = \cos \cos x$;     (j) $f(x) = 2^{x^2 + 2(\cos 2^x)^2}$.

3.2   Let $X = \mathbb{R} \setminus \{0, 1\}$. Define $f_i : X \to X$ for $1 \leqslant i \leqslant 6$ by

$$f_1(x) = x, \quad f_2(x) = 1 - x, \quad f_3(x) = \frac{x-1}{x},$$

$$f_4(x) = \frac{1}{x}, \quad f_5(x) = \frac{1}{1-x}, \quad f_6(x) = \frac{x}{x-1}.$$

Show that $f_i \circ f_j \in \{f_k \mid 1 \leqslant k \leqslant 6\}$ for all $i, j$.

3.3 Let $f : \mathbb{R} \to \mathbb{R}$ be defined by

$$f(x) = [1 + (1 - x)^3]^{1/3}.$$

Express $f$ as a composite of four maps none of which is the identity.

3.4 Let $f : X \to X$ and define a relation $R$ on $X$ by $xRy$ if and only if $y = f(x)$. Prove that

(a) $R$ is reflexive if and only if $f = \mathrm{id}_X$;
(b) $R$ is symmetric if and only if $f^2 = \mathrm{id}_X$;
(c) $R$ is transitive if and only if $f^2 = f$.

3.5 If $f : X \to X$ and if $g$ is defined by

$$g = \{(y, x) \in X \times X \mid y = f(x)\}$$

prove that $g \circ f$ is an equivalence relation, where $g \circ f$ denotes the product relation defined in 2.8.

3.6 If $f : A \to B$ prove that the relation $R_f$ defined on $A$ by

$$xR_f y \Leftrightarrow f(x) = f(y)$$

is an equivalence relation.
Let $f : \mathbb{R} \times \mathbb{R} \to \mathbb{R} \times \mathbb{R}$ be defined by

$$f(x, y) = \begin{cases} \left( \dfrac{x}{\sqrt{(x^2 + y^2)}}, \dfrac{y}{\sqrt{(x^2 + y^2)}} \right) & \text{if} \quad (x, y) \neq (0, 0); \\ (0, 0) & \text{if} \quad (x, y) = (0, 0). \end{cases}$$

Find the $R_f$-class of $(1, 0)$. Describe geometrically the $R_f$-class of $(a, b)$.

3.7 For each of the following relations on $\mathbb{R}$

(i) sketch its graph;
(ii) find its domain and image;
(iii) say whether or not it is a mapping.

(a) $x^2 \leqslant y$;   (b) $\sin x = y$;   (c) $x = \sin y$;
(d) $x - 2 \leqslant y \leqslant x + 1$;   (e) $y = |x|$;   (f) $x + y = 1$;
(g) $|x| + y = 1$;   (h) $x + |y| = 1$;   (i) $|x| + |y| = 1$;
(j) $|y| \leqslant |x| \leqslant 1$;   (k) $y = |x| - [\![x]\!]$.

15

**3.8** Find the domain and image of the following relations on $\mathbb{R}$. Are any of them mappings?

      (a) $\{(x,y) \mid x^2 + 4y^2 = 1\}$;

      (b) $\{(x,y) \mid x^2 = y^2\}$;

      (c) $\{(x,y) \mid y \geqslant 0, y \leqslant x, x + y \leqslant 1\}$;

      (d) $\{(x,y) \mid x^2 + y^2 \leqslant 1, x \geqslant 0\}$;

      (e) $\{(x,y) \mid y = 2x - 1\}$.

**3.9** Let $f$ be the subset of $\mathbb{Q} \times \mathbb{Z}$ given by

$$f = \{(x,y) \mid y \text{ is the least integer with } y \geqslant x\}.$$

Determine the domain and image of $f$. Is $f$ a mapping? Answer the same question for the subset of $\mathbb{Z} \times \mathbb{Q}$ given by

$$f = \{(x,y) \mid x \text{ is the least integer with } x \geqslant y\}.$$

**3.10** Sketch the function $f : \mathbb{R} \to \mathbb{R}$ given by

$$f(x) = ||x - 1| - 1|.$$

**3.11** If $f : \mathbb{R} \to \mathbb{R}$ is defined by

$$f(x) = 2x^2 + 6x + 7$$

determine the set $\{x \in \mathbb{R} \mid f(x) \leqslant x + 5\}$.

**3.12** Let $f : \mathbb{R} \to \mathbb{R}$. Show that there exist mappings $g, h : \mathbb{R} \to \mathbb{R}$ such that $f = g + h$ with $g(x) = g(-x)$ and $h(x) = -h(-x)$ for all $x \in \mathbb{R}$.

**3.13** Let $f, g : \mathbb{R} \to \mathbb{R}$ be given by

$$f(x) = (x + 1)^2, \quad g(x) = 2x - 1.$$

Determine the mappings $f \circ g, g \circ f$ and the set

$$\{x \in \mathbb{R} \mid (f \circ g)(x) = (g \circ f)(x)\}.$$

Given the subsets of $\mathbb{R} \times \mathbb{R}$ described by

$$A = \{(x,y) \mid x \geqslant 0, y \geqslant 0\};$$
$$B = \{(x,y) \mid y \leqslant f(x)\};$$
$$C = \{(x,y) \mid y \geqslant g(x)\};$$
$$D = \{(x,y) \mid y \leqslant 1 - x\},$$

describe the set $A \cap B \cap C \cap D$.

**3.14** Sketch the subset $S$ of $\mathbb{R} \times \mathbb{R}$ given by

$$S = \{(x,y) \in \mathbb{R} \times \mathbb{R} \mid |x| + |y| = 1\}.$$

If the mapping $d : S \to \mathbb{R}$ is given by
$$d(x,y) = \sqrt{(x^2 + y^2)}$$
determine Im $d$.

**3.15** Sketch the subset $S$ of $\mathbb{R} \times \mathbb{R}$ given by
$$S = \{(x,y) \mid |x| + |y| \geqslant 1 \quad \text{and} \quad x^2 + y^2 \leqslant 1\}.$$
If the mapping $f : S \to \mathbb{R}$ is given by
$$f(x,y) = \sqrt{(x^2 + y^2)}$$
determine Im $f$.

**3.16** Let $S = \{(x, y) \in \mathbb{R} \times \mathbb{R} \mid x^2 + y^2 - 10x + 16 = 0\}$ and let $f : S \to \mathbb{R}$ be given by
$$f(x,y) = \frac{y}{x}.$$
By considering this mapping geometrically, determine Im $f$.

**3.17** Let $S = \{(x, y) \in \mathbb{R} \times \mathbb{R} \mid x^2 + y^2 - 6x - 8y + 21 = 0\}$ and let $f : S \to \mathbb{R}$ be given by
$$f(x,y) = \sqrt{(x^2 + y^2)}.$$
By considering this mapping geometrically, determine Im $f$.

**3.18** Each of the following describes a mapping $f : A \to B$ by exhibiting the set $A$, the set $B$, and the image $f(x)$ for every $x \in A$. Determine which of the mappings are injective or surjective.

(a) $A = \mathbb{Z}, \quad B = \mathbb{Q}, \quad f(x) = 2x + 1;$

(b) $A = \mathbb{Z}, \quad B = \mathbb{Z} \times \mathbb{Z}, \quad f(x) = (x - 1, 1);$

(c) $A = \mathbb{R}, \quad B = \mathbb{R}, \quad f(x) = x^3;$

(d) $A = \mathbb{C}, \quad B = \mathbb{R}, \quad f(x) = |x|;$

(e) $A = \mathbb{R}, \quad B = \mathbb{R}, \quad f(x) = x \sin x;$

(f) $A = \mathbb{R}, \quad B = \mathbb{R}, \quad f(x) = (x - 1)x(x + 1);$

(g) $A = \mathbb{R}, \quad B = \mathbb{R}, \quad f(x) = x^2 + x + 1.$

**3.19** If $f : X \to Y$ and $A, B$ are subsets of $X$ prove that

(a) $f(A \cap B) \subseteq f(A) \cap f(B);$

(b) $f(A \cup B) = f(A) \cup f(B).$

Prove also that

(c) $f$ is injective if and only if $f(A \cap B) = f(A) \cap f(B)$ for all $A, B$;

(d) $f$ is surjective if and only if $f(X \setminus A) \supseteq Y \setminus f(A)$ for all $A$;

(e) $f$ is bijective if and only if $f(X \setminus A) = Y \setminus f(A)$ for all $A$.

**3.20** If $X$ and $Y$ are sets denote by $Y^X$ the set of all mappings from $X$ to $Y$. Show that, for all sets $A, B$ and $C$, it is possible to find a bijection between

    (*a*) $(A \times B)^C$ and $A^C \times B^C$;

    (*b*) $(A^B)^C$ and $A^{B \times C}$;

    (*c*) $A^{B \cup C}$ and $A^B \times A^C$ if $B \cap C = \emptyset$.

If there is a bijection between $A^{B \cup C}$ and $A^B \times A^C$ is it necessary for $B \cap C$ to be empty?

**3.21** Describe explicitly a bijection

    (*a*) from $[0, 1]$ to $[1, 2]$;

    (*b*) from $[0, 1]$ to $[0, 2]$;

    (*c*) from $]-1, 1[$ to $\mathbb{R}$;

    (*d*) from $[0, 1]$ to $[0, 1[$;

    (*e*) from $[-1, 1]$ to $\mathbb{R}$.

(*Hint:* for (*d*) consider a map that acts as the identity map except on some set of rationals.)

**3.22** The functions $f : \mathbb{R} \to \mathbb{R}$ and $g : \mathbb{R} \to \mathbb{R}$ are defined by

$$f(x) = \begin{cases} 4x + 1 & \text{if } x \geqslant 0; \\ x & \text{if } x < 0, \end{cases}$$

$$g(x) = \begin{cases} 3x & \text{if } x \geqslant 0; \\ x + 3 & \text{if } x < 0. \end{cases}$$

Show that $g \circ f$ is a bijection and give a formula for $(g \circ f)^{-1}$. Show also that $f \circ g$ is neither injective nor surjective.

**3.23** The functions $f : \mathbb{R} \to \mathbb{R}$ and $g : \mathbb{R} \to \mathbb{R}$ are defined by

$$f(x) = \begin{cases} 1 - x & \text{if } x \geqslant 0; \\ x^2 & \text{if } x < 0, \end{cases}$$

$$g(x) = \begin{cases} x & \text{if } x \geqslant 0; \\ x - 1 & \text{if } x < 0. \end{cases}$$

Find a formula for $f \circ g$ and draw its graph. Show that $f \circ g$ is a bijection and find its inverse.

    Find also a formula for $g \circ f$ and draw its graph. Show that $g \circ f$ is neither injective nor surjective.

**3.24** Let $f : \mathbb{R} \to \mathbb{R}$ be given by

$$f(x) = x \, |x|.$$

18

Show that $f$ is a bijection and determine $f^{-1}$. Is the mapping $g : \mathbb{R} \to \mathbb{R}$ given by

$$g(x) = x^2 \, |x|$$

a bijection? If so, determine $g^{-1}$.

3.25  Using calculus, or otherwise, prove that the mapping $f : \mathbb{R} \to \mathbb{R}$ described by
$$f(x) = x^3 + ax^2 + bx + c \quad (a, b, c \in \mathbb{R})$$
is a bijection if and only if $a^2 \leqslant 3b$.

3.26  Let $\mathbb{R}_+ = \{x \in \mathbb{R} \mid x \geqslant 0\}$ and let $f : \mathbb{R}_+ \to \mathbb{R}_+$ be given by
$$f(x) = \frac{1}{1 + x^2}.$$
Prove that $f$ is injective and find two distinct mappings $g, h : \mathbb{R}_+ \to \mathbb{R}_+$ such that $g \circ f = h \circ f = \mathrm{id}_{\mathbb{R}_+}$.

3.27  Let $f : \mathbb{R} \to \mathbb{R}$ be defined by
$$f(x) = \cos\,[\pi(x - [\![x]\!])] - 2[\![x]\!].$$
Prove that $f$ is injective and find two distinct mappings $g, h : \mathbb{R}_+ \to \mathbb{R}_+$ such a bijection?

3.28  Prove that if $\mathbb{N}^* = \mathbb{N} \setminus \{0\}$ then the mapping $f : \mathbb{N}^* \to \mathbb{Z}$ given by
$$f(n) = (-1)^n \left[\!\left[ \frac{n}{2} \right]\!\right]$$
is a bijection and obtain a formula for $f^{-1}$.

3.29  Let $g : \mathbb{R} \to \mathbb{R}$ be given by
$$g(x) = 3 + 4x.$$
Prove by induction that, for all positive integers $n$,
$$g^n(x) = (4^n - 1) + 4^n x.$$
If for every positive integer $k$ we interpret $g^{-k}$ as the inverse of the function $g^k$, prove that the above formula holds also for all negative integers $n$.

3.30  Use induction to prove that if $A_1, \ldots, A_n$ are subsets of a set $E$ then

$$\left| \bigcup_{i=1}^{n} A_i \right| = \sum_{i=1}^{n} |A_i| - \sum_{i<j} |A_i \cap A_j|$$
$$+ \sum_{i<j<k} |A_i \cap A_j \cap A_k| - \ldots + (-1)^n \left| \bigcap_{i=1}^{n} A_i \right|.$$

Let $A$ and $B$ be finite sets with $|A| = m$ and $|B| = n$. How many mappings are there from $A$ to $B$? How many injections are there from $A$ to $B$? Show that there are

$$\sum_{i=0}^{n} (-1)^i \binom{n}{i} (n-i)^m$$

surjections from $A$ to $B$.

**3.31**   Give an example of non-empty sets $A$, $B$ and $C$ with the property that there are injections

$$A \rightarrow B \rightarrow C \rightarrow A$$

none of which are bijections.

**3.32**   Consider the mapping $f : \mathbb{R} \rightarrow \mathbb{R}$ given by

$$f(x) = \frac{4x}{(x^2 + 1)}.$$

Sketch the graph of $f$. Find an interval $A = [-k, k]$ on the $x$-axis such that
   (a) $\{f(x) \mid x \in A\} = \mathrm{Im}\, f$;
   (b) $g : A \rightarrow \mathrm{Im}\, f$ given by $g(a) = f(a)$ for every $a \in A$ is a bijection.
Obtain a formula for $g^{-1} : \mathrm{Im}\, f \rightarrow A$.

**3.33**   Sketch the graph of the function $f : \mathbb{R} \rightarrow \mathbb{R}$ given by

$$f(x) = 3 + 2x - x^2.$$

Show that $f$ is not injective. Determine $\mathrm{Im}\, f$ and find a subset $A$ of $\mathbb{R}$ such that the restriction of $f$ to $A$ induces a bijection $g : A \rightarrow \mathrm{Im}\, f$. Obtain a formula for the inverse of this bijection.

**3.34**   Let $p$ be a fixed positive integer. Prove that the mapping $f : \mathbb{Z} \rightarrow \mathbb{Z}$ given by

$$f(n) = \begin{cases} n + p & \text{if } n \text{ is divisible by } p, \\ n & \text{if } n \text{ is not divisible by } p, \end{cases}$$

is a bijection, and determine $f^{-1}$.

**3.35**   Prove that the mapping $f : \mathbb{R} \rightarrow \mathbb{R}$ given by

$$f(x) = \begin{cases} x^4 & \text{if } x \geqslant 0; \\ x(2 - x) & \text{if } x < 0, \end{cases}$$

is a bijection, and find its inverse.

20

**3.36** Define $f : \mathbb{N} \to \mathbb{N}$ by $f(0) = 0$, $f(1) = 1$, and $f(n) = f(n-1) + f(n-2)$ for $n \geqslant 2$. Prove that

      (a) $f(i) < f(i+1)$ for all $i \geqslant 2$;

      (b) there exist precisely four $i \in \mathbb{N}$ with $(f \circ f)(i) = f(i)$.

Writing $f_n$ for $f(n)$, prove also that

      (c) $f_{5n}$ is divisible by 5;

      (d) $f_{n+2} = 1 + \Sigma_{i=1}^{n} f_i$;

      (e) $f_{n-1} f_{n+1} - f_n^2 = (-1)^n$ for $n \geqslant 1$;

      (f) $2^n f_n = [(1 + \sqrt{5})^n - (1 - \sqrt{5})^n]/\sqrt{5}$.

**3.37** Let $X = \{1, 2, 3, 4\}$ and define $f : X \to X$ by

$$\begin{cases} f(x) = x + 1 & \text{if } x \leqslant 3, \\ f(4) = 1. \end{cases}$$

Show that there is only one mapping $g : X \to X$ with the property that $g(1) = 3$ and $f \circ g = g \circ f$. Find $g$. Is it true that there is only one mapping $h : X \to X$ with $h(1) = 1$ and $f \circ h = h \circ f$?

**3.38**   (a) If $\alpha : \mathbb{N} \to \mathbb{N}$ is given by $\alpha(n) = n + 1$ show that there is no mapping $g : \mathbb{N} \to \mathbb{N}$ such that $\alpha \circ g = \mathrm{id}_{\mathbb{N}}$ but that there are infinitely many mappings $k : \mathbb{N} \to \mathbb{N}$ such that $k \circ \alpha = \mathrm{id}_{\mathbb{N}}$.

  (b) If $\beta : \mathbb{N} \to \mathbb{N}$ is given by

$$\beta(n) = \begin{cases} n/2 & \text{if } n \text{ is even}; \\ (n-1)/2 & \text{if } n \text{ is odd}, \end{cases}$$

show that there is no mapping $f : \mathbb{N} \to \mathbb{N}$ such that $f \circ \beta = \mathrm{id}_{\mathbb{N}}$ but that there are infinitely many mappings $k : \mathbb{N} \to \mathbb{N}$ such that $\beta \circ k = \mathrm{id}_{\mathbb{N}}$.

**3.39** Let $A$ be a set and let $f : A \to \mathbf{P}(A)$. Define the subset $X$ of $A$ by $X = \{a \in A \mid a \notin f(a)\}$. Can there exist $a \in A$ with $f(x) = X$? Can $f$ be a surjection? Can $f$ be an injection?

**3.40** Let $\mathbb{N}^* = \{1, 2, 3, \ldots\}$ and for every $k \in \mathbb{N}^*$ define

$$I_k = \{x \in \mathbb{N}^* \mid \tfrac{1}{2}k(k-1) < x \leqslant \tfrac{1}{2}k(k+1)\}.$$

  (a) Show that $I_k$ has $k$ elements and that $\{I_k \mid k \in \mathbb{N}^*\}$ is a partition of $\mathbb{N}^*$.

  (b) Define $f : \mathbb{N}^* \times \mathbb{N}^* \to \mathbb{N}^*$ by

$$f(m, n) = \tfrac{1}{2}(m + n - 2)(m + n - 1) + m.$$

Show that $f(m, n) \in I_{m+n-1}$ and deduce that

$$f(p,q) \in I_{m+n-1} \Leftrightarrow p+q = m+n.$$

Hence show that $f$ is injective.

(c) For $1 \leqslant r \leqslant k$ show that $f(r, k+1-r) \in I_k$ and deduce that $f$ is also surjective.

**3.41** Let $S = \mathbb{R} \setminus \{1, -1\}$. Find a mapping $f : S \to S$ such that $f \circ f = -\mathrm{id}_S$. (*Hint:* try mapping $]-1, 1[$ to its complement in $S$.)

**3.42** Given mappings $f : A \to B$, $g : B \to C$, $h : C \to D$, suppose that $g \circ f$ and $h \circ g$ are bijections. Prove that $f, g, h$ are all bijections.

**3.43** Let $\mathbb{Q}_+ = \{x \in \mathbb{Q} \mid x \geqslant 0\}$. If $a/b, c/d \in \mathbb{Q}_+$ prove that

$$\frac{a}{b} = \frac{c}{d} \Rightarrow \left| \frac{a+b}{\mathrm{hcf}(a,b)} \right| = \left| \frac{c+d}{\mathrm{hcf}(c,d)} \right|.$$

Deduce that the prescription

$$f\left(\frac{a}{b}\right) = \left| \frac{a+b}{\mathrm{hcf}(a,b)} \right|$$

defines a mapping $f : \mathbb{Q}_+ \to \mathbb{Q}_+$. Is $f$ a bijection?

**3.44** For mappings $f, g : \mathbb{R} \to \mathbb{R}$ and every $\lambda \in \mathbb{R}$ define the mappings $f + g$, $f \cdot g$ and $\lambda f$ from $\mathbb{R}$ to $\mathbb{R}$ in the usual way, namely by setting

$$(f+g)(x) = f(x) + g(x), \quad (f \cdot g)(x) = f(x)g(x),$$
$$(\lambda f)(x) = \lambda f(x)$$

for every $x \in \mathbb{R}$.

(a) Show that there are bijections $f$, $g$ with $f + g$ not a bijection. Show also that there are bijections $f$, $g$ with $f \cdot g$ not a bijection. Do there exist bijections $f, g$ such that neither $f + g$ nor $f \cdot g$ is a bijection?

(b) Prove that if $\lambda \neq 0$ then $\lambda f$ is a bijection if and only if $f$ is a bijection.

(c) Define $[fg] : \mathbb{R} \to \mathbb{R}$ by $[f, g] = f \circ g - g \circ f$. Do there exist bijections $f, g$ with $[fg]$ a bijection?

(d) If $0$ denotes the mapping from $\mathbb{R}$ to $\mathbb{R}$ described by $x \to 0$, prove that, for all mappings $f, g, h : \mathbb{R} \to \mathbb{R}$,

$$[[fg]h] + [[gh]f] + [[hf]g] = 0.$$

**3.45** (a) Define a relation $R$ on $\mathbb{C}^* = \mathbb{C} \setminus \{0\}$ by

$$(a+ib)R(c+id) \Leftrightarrow a\sqrt{(c^2+d^2)} = c\sqrt{(a^2+b^2)}.$$

Show that $R$ is an equivalence relation and describe geometrically the $R$-classes.

(b) Let $U = \{[x]_R \mid x \in \mathbb{C}^*\}$ be the set of $R$-classes. Show that if $xRy$

then $x^2 R y^2$ and deduce that the relation $f$ on $U$ given by

$$f = \{([x]_R, [x^2]_R) \mid x \in \mathbb{C}^*\}$$

is a mapping. Is $f$ injective? Is $f$ surjective?

(c) For each of the following relations $f$ on $U$ determine whether $f$ is a mapping $U \to U$. For those relations which are mappings, determine which are injective and which are surjective.

(i) $f = \{([x]_R, [2x]_R) \mid x \in \mathbb{C}^*\}$;

(ii) $f = \{([x]_R, [x+2]_R) \mid x \in \mathbb{C}^*\}$;

(iii) $f = \{([x]_R, [x^{-1}]_R) \mid x \in \mathbb{C}^*\}$.

**3.46** (a) Define a mapping $f : \mathbb{N} \to \mathbb{N}$ by

$$f(n) = \text{the sum of the digits of } n.$$

Is $f$ injective? Is $f$ surjective?

(b) For every $i \geqslant 1$ let $R_i$ be the equivalence relation on $\mathbb{N}$ defined by

$$x R_i y \Leftrightarrow f^i(x) = f^i(y).$$

Describe the class $[1]_{R_1}$. Prove that if $1 \leqslant i \leqslant j$ then $[1]_{R_i} \subseteq [1]_{R_j}$. Is it true that $[n]_{R_i} \subseteq [n]_{R_j}$ for every $n \in \mathbb{N}$? Is $[1]_{R_1} = [1]_{R_2}$?

(c) Prove that $n$ is divisible by 9 if and only if $n \in [9]_{R_i}$ for some $i > 0$. Find a similar criterion for $n$ to be divisible by 3.

(d) Define a relation $R$ on $\mathbb{N}$ by

$$x R y \Leftrightarrow x R_i y \text{ for some } i > 0.$$

Prove that $R$ is an equivalence relation. Show that there are infinitely many $R_i$-classes for each $i > 0$ but that there are only finitely many $R$-classes. How many $R$-classes are there?

# Solutions to Chapter 1

*1.1*    Each of the statements is true. Since $\emptyset$ is a subset of every set we have $\emptyset \subseteq A$. From the definition of $A$ we see that $\emptyset \in A$ and $\{\emptyset\} \in A$, and these give, respectively, $\{\emptyset\} \subseteq A$ and $\{\{\emptyset\}\} \subseteq A$. Since $\emptyset$ and $\{\emptyset\}$ are elements of $A$ it follows that $\{\{\emptyset\}, \emptyset\} \subseteq A$. Finally, $\{\{\emptyset\}, \emptyset\} = \{\emptyset, \{\emptyset\}\}$ is, by definition, an element of $A$.

*1.2*    The only subset of $\emptyset$ is $\emptyset$ and so $\mathbf{P}(\emptyset) = \{\emptyset\}$. It follows that $\mathbf{P}(\mathbf{P}(\emptyset)) = \{\emptyset, \{\emptyset\}\}$. Finally,

$$\mathbf{P}(\mathbf{P}(\mathbf{P}(\emptyset))) = \{\emptyset, \{\emptyset\}, \{\{\emptyset\}\}, \{\emptyset, \{\emptyset\}\}\}.$$

*1.3*    The subsets of $E = \{1, \{1\}, 2, \{1, 2\}\}$ are

$$\emptyset, \ \{1\}, \ \{\{1\}\}, \ \{2\}, \ \{\{1, 2\}\}, \ \{1, \{1\}\}, \ \{1, 2\}, \ \{1, \{1, 2\}\},$$
$$\{\{1\}, 2\}, \ \{\{1\}, \{1, 2\}\}, \ \{2, \{1, 2\}\}, \ \{1, \{1\}, 2\}, \ \{\{1\}, 2, \{1, 2\}\},$$
$$\{1, \{1\}, \{1, 2\}\}, \ \{1, 2, \{1, 2\}\}, \ E.$$

From this list we see that $E \cap \mathbf{P}(E) = \{\{1\}, \{1, 2\}\}$.

*1.4*    Four examples are

$$\emptyset, \ \ \{\emptyset\}, \ \ \{\emptyset, \{\emptyset\}\}, \ \ \{\emptyset, \{\emptyset\}, \{\emptyset, \{\emptyset\}\}\}.$$

*1.5*    Yes. For example, take $A = \emptyset, B = \{\emptyset\}$ and $C = \{\emptyset, \{\emptyset\}\}$.

*1.6*        (a) False; consider, for example, $A = \{1\}, B = \{2\}$, and $C = \{\{1\}\}$.
        (b) False; consider $A = \{1\}, B = \{2\}, C = A$.
        (c) False; consider $A = \{1\}, B = \{\{1\}, 1\}, C = \{\{1\}, 2\}$.
        (d) False; consider $A = B = C = \{1\}$.
        (e) False; consider $A = \{1\}, B = \{1, 2\}, C = \{\{1, 2\}, \{1\}\}$.
        (f) False; consider $A = \{1, 2\}, B = \{2, 3\}, C = \{2, 4\}$.
        (g) False; consider $A = \{1\}, B = \{\emptyset\}, C = \{2\}$.

*Solutions to Chapter 1*

(*h*) True; using the distributive law we have

$$(A \cup B) \cap C = (A \cap C) \cup (B \cap C) = \emptyset \cup \emptyset = \emptyset.$$

1.7    Take $x = z \neq y$.

1.8    (*a*) Using the de Morgan law we have

$$(A \cup (A \cup B)')')' = A' \cap (A \cup B)''$$
$$= A' \cap (A \cup B)$$
$$= (A' \cap A) \cup (A' \cap B)$$
$$= \emptyset \cup (A' \cap B)$$
$$= A' \cap B.$$

(*b*) The expression clearly reduces to

$$(A \cap (B \cup A') \cap X)' = (A \cap (B \cup A'))'$$
$$= A' \cup (B \cup A')'$$
$$= A' \cup (B' \cap A)$$
$$= (A' \cup B') \cap (A' \cup A)$$
$$= (A' \cup B') \cap X$$
$$= A' \cup B'.$$

(*c*) Since $(B \cap C) \cup C = C$, the expression reduces to

$$(A \cup (B' \cap C') \cup C)' = ((A \cup B' \cup C) \cap (A \cup C' \cup C))'$$
$$= ((A \cup B' \cup C) \cap X)'$$
$$= (A \cup B' \cup C)'$$
$$= A' \cap B \cap C'.$$

1.9    The expression $(A \cap B') \cup (A' \cap B)$ is none other than $A \triangle B$. The Venn diagram for $A \triangle B$ is as shown in Fig. S1.1.

**Fig.S1.1**

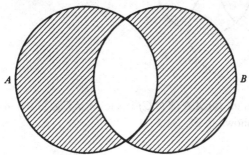

25

It is clear from this that $A \triangle B = A \cup B$ if and only if $A \cap B = \emptyset$.

*1.10*          (*a*) Note that another expression for $A \triangle B$ is $(A \cup B) \setminus (A \cap B)$. Using this, we see that the Venn diagrams for $A \triangle (B \triangle C)$ and $(A \triangle B) \triangle C$ are the same: see Fig. S1.2.

**Fig.S1.2**

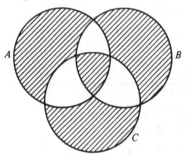

(*b*) Using the fact that $A \triangle B = (A \cup B) \setminus (A \cap B)$ we have

$$A \triangle B \triangle (A \cap B) = ((A \triangle B) \cup (A \cap B)) \setminus ((A \triangle B) \cap (A \cap B))$$
$$= (A \cup B) \setminus \emptyset$$
$$= A \cup B.$$

(*c*) This follows from the fact that the Venn diagram for $A \cap (B \triangle C)$ is the same as that for $(A \cap B) \triangle (A \cap C)$: see Fig. S1.3.

**Fig.S1.3**

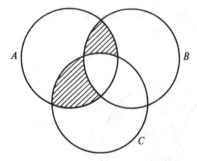

(*d*) The Venn diagram for $A \triangle (B \cap C)$ is shown in Fig. S1.4 and that for $(A \triangle B) \cap (A \triangle C)$ is shown in Fig. S1.5.

## Solutions to Chapter 1

**Fig.S1.4**

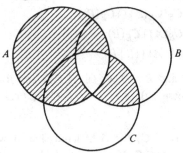

**Fig.S1.5**

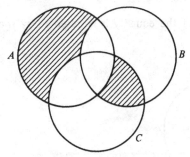

These are the same if and only if

$$A \cap B \cap C' = \emptyset = A \cap B' \cap C,$$

i.e. if and only if $A \cap B \subseteq C$ and $A \cap C \subseteq B$, which is the case if and only if $A \cap B = A \cap C$.

*1.11*   We have

$$A \mid A = C_E(A \cap A) = C_E(A).$$

Also,

$$\begin{aligned}(A \mid B) \mid (A \mid B) &= C_E(A \cap B) \mid C_E(A \cap B) \\ &= C_E(C_E(A \cap B) \cap C_E(A \cap B)) \\ &= C_E C_E(A \cap B) \\ &= A \cap B,\end{aligned}$$

27

and

$$A \cup B = C_E C_E(A \cup B) = C_E(C_E(A) \cap C_E(B))$$
$$= C_E(A) \mid C_E(B)$$
$$= (A \mid A) \mid (B \mid B).$$

*1.12*   $\{A, B\} = \{C, D\}$ implies (i) $A = C$ and $B = D$, or (ii) $A = D$ and $B = C$. In case (i), $A \cap B = C \cap D$; and in case (ii), $A \cap B = D \cap C = C \cap D$. Argue similarly for $\cup$.

*1.13*   We have $A \cap (B \cup C) = (A \cap B) \cup (A \cap C)$ the right hand side of which is contained in $(A \cap B) \cup C$ since $A \cap C \subseteq C$. If now $C \subseteq A$ then $A \cap C = C$ and we have equality in the above. Conversely, if the equality $A \cap (B \cup C) = (A \cap B) \cup C$ holds then since the left hand side is contained in $A$ and the right hand side contains $C$ we have that $C \subseteq A$.

*1.14*   As is readily verified, each side of the equality is represented by the Venn diagram shown in Fig. S1.6.

**Fig.S1.6**

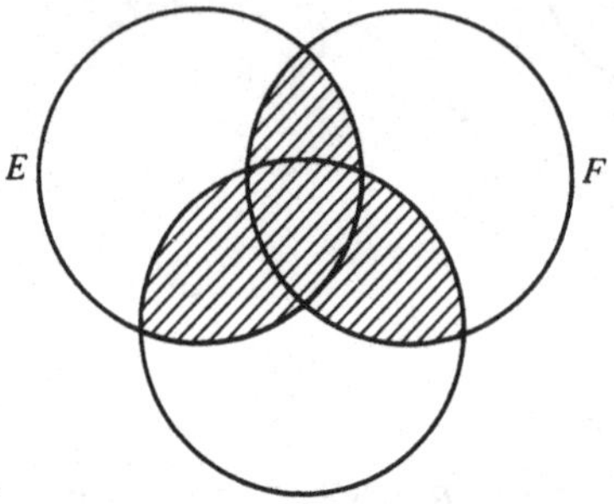

*1.15*   Take, for example, $A = C = \{1\}, B = D = \{2\}$. We have
$$(A \cup B) \times (C \cup D) = \{(1, 1), (1, 2), (2, 1), (2, 2)\},$$
$$(A \times C) \cup (B \times D) = \{(1, 1), (2, 2)\}.$$

*1.16*   Suppose first that $y \neq x$ and $y^* \neq x^*$. Then we have
$$(x, y) = (x^*, y^*) \Leftrightarrow \{\{x\}, \{x, y\}\} = \{\{x^*\}, \{x^*, y^*\}\}$$
$$\Leftrightarrow \{x\} = \{x^*\} \text{ and } \{x, y\} = \{x^*, y^*\}$$
$$\Leftrightarrow x = x^* \text{ and } y = y^*,$$

so the result holds in this case.

28

*Solutions to Chapter 1*

Suppose now that $y = x$. Then we have
$$(x, y) = (x, x) = \{\{x\}, \{x, x\}\} = \{\{x\}, \{x\}\} = \{\{x\}\}$$
and so in this case
$$(x, y) = (x^*, y^*) \Leftrightarrow \{\{x\}\} = \{\{x^*\}, \{x^*, y^*\}\}$$
$$\Leftrightarrow \{x\} = \{x^*\} = \{x^*, y^*\}$$
$$\Leftrightarrow x^* = y^* = x(=y).$$

This establishes the result in this case. The only other case to consider is that in which $y^* = x^*$, and this is similar to the case in which $y = x$.

Finally, we have
$$\{x\} \times \{x\} = \{(x, x)\} = \{\{\{x\}, \{x, x\}\}\} = \{\{\{x\}\}\}.$$

1.17   If $A = B = C$ then clearly $(A \times B) \cup (B \times A) = C \times C$. Conversely, suppose that this equality holds and let $a \in A$ and $b \in B$. Since $(a, b)$ belongs to the left hand side, it belongs to the right hand side, so $a \in C$ and $b \in C$. Thus $A \subseteq C$ and $B \subseteq C$. However, if $x \in C$ then $(x, x) \in C \times C$, so either $(x, x) \in A \times B$ or $(x, x) \in B \times A$. In either case, $x \in A$ and $x \in B$ from which we conclude that $C = A$ and $C = B$.

1.18   Observe that
$$X \cap Y = X \setminus (X \setminus Y).$$

1.19   If $X, Y \in \mathscr{F}$ then $(X \setminus Y) \setminus (Y \setminus X) \in (\mathscr{F}^0)^0$. But clearly
$$(X \setminus Y) \setminus (Y \setminus X) = X \setminus Y.$$
Thus $\mathscr{F}^0 \subseteq (\mathscr{F}^0)^0$.

Consider now $\mathscr{F} = \{X, Y, Z\}$ where $X = \{1, 2\}$, $Y = \{2, 3, 4\}$, and $Z = \{4\}$. We have
$$\mathscr{F}^0 = \{\{1\}, \{1, 2\}, \{4\}, \{2, 3\}, \{3, 4\}, \emptyset\}.$$
Since $(\mathscr{F}^0)^0$ contains $\{2\}$ and $\mathscr{F}^0$ does not, the desired inequality follows.

1.20   (a) True; for
$$X \in P(A) \cap P(B) \Leftrightarrow X \subseteq A \text{ and } X \subseteq B$$
$$\Leftrightarrow X \subseteq A \cap B$$
$$\Leftrightarrow X \in P(A \cap B).$$

(b) False. For example, take $A = \{1\}$, $B = \{2\}$. Then we have $\{1, 2\} \in P(A \cup B)$ but
$$\{1, 2\} \notin P(A) \cup P(B) = \{\emptyset, \{1\}, \{2\}\}.$$

29

*1.21*    The Venn diagrams for $A \setminus (B \cup C)$ and $(A \setminus B) \cup (A \setminus C)$ are shown in Figs. S1.7 and S1.8, respectively.

**Fig.S1.7**

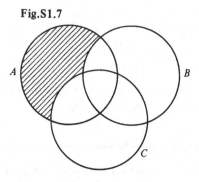

**Fig.S1.8**

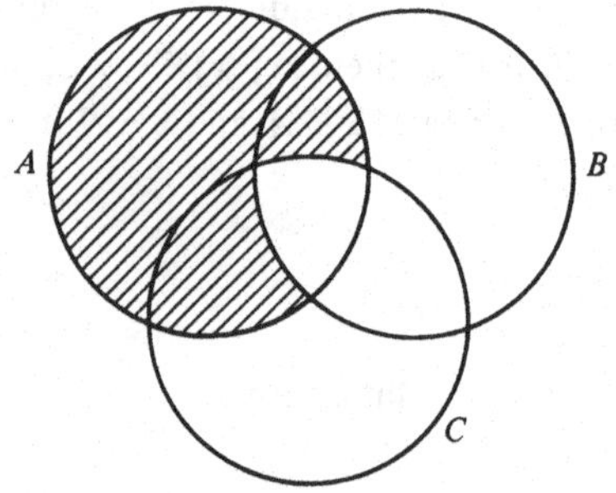

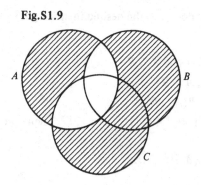

Clearly, we have equality if and only if $A \cap (B \bigtriangleup C) = \emptyset$.

As for $A \bigtriangleup (B \cup C)$ and $(A \bigtriangleup B) \cup (A \bigtriangleup C)$, the respective Venn diagrams are shown in Figs. S1.9 and S1.10.

**Fig.S1.9**

Fig.S1.10

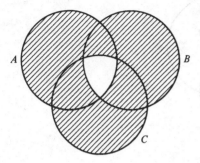

Again we have equality if and only if $A \cap (B \triangle C) = \emptyset$; hence the result.

Consider now $A = \{1\}, B = C = \{2\}$. We have

$$A \triangle (B \cup C) = \{1\} \triangle \{2\} = \{1, 2\}$$

and

$$(A \triangle B) \cup (A \triangle C) = \{1, 2\} \cup \{1, 2\} = \{1, 2\}.$$

However, taking $A = B = \{1\}, C = \{2\}$ we have

$$A \triangle (B \cup C) = \{1\} \triangle \{1, 2\} = \{2\}$$

whereas

$$(A \triangle B) \cup (A \triangle C) = \emptyset \cup \{1, 2\} = \{1, 2\}.$$

1.22 Assume, without loss of generality, that $i < j$. From the definition we have $B_i \subseteq A_i$ and $B_j = A_j \setminus \bigcup_{k=1}^{j-1} A_k$. Now $B_i \subseteq A_i \subseteq \bigcup_{k=1}^{j-1} A_k$ so $B_i \not\subseteq B_j$ and it follows that $B_i \cap B_j = \emptyset$.

Since $B_i \subseteq A_i$ for every $i$, we have $\bigcup_{i=1}^{m} B_i \subseteq \bigcup_{i=1}^{m} A_i$. Now let $x \in \bigcup_{i=1}^{m} A_i$. Let $t$ be the least integer such that $x \in A_t$ and $x \notin A_i$ for $i < t$. Then $x \in B_t$ and so $x \in \bigcup_{i=1}^{m} B_i$, showing that $\bigcup_{i=1}^{m} A_i \subseteq \bigcup_{i=1}^{m} B_i$.

1.23 Consider the subset of $B$ given by $X = B \setminus A$. Clearly, $X \cap A = \emptyset$ and $X \cup A = B$, so a subset with the given properties exists. To show that it is unique, let $Y$ be a subset of $B$ with these properties. Since $Y \cap A = \emptyset$ we have that $Y \subseteq B \setminus A$. But $Y \supseteq (Y \cup A) \setminus A = B \setminus A$ whence we have that $Y = X$.

1.24 Draw a Venn diagram to illustrate the given information, putting numbers in the appropriate places to indicate the failures (see Fig. S1.11).

**Fig.S1.11**

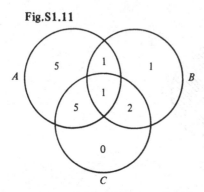

From the diagram we see that the number of students who failed in any of the subjects is 15. Consequently $41 - 15 = 26$ passed in all three subjects.

1.25    Let $A$, $C$, $G$, $T$ signify algebra, calculus, geometry and trigonometry respectively, and let $|X|$ denote the number of students who study subject $X$. Assume, without loss of generality, that there are 100 students in the class. Then from the fact that

$$|A \text{ or } B| = |A| + |B| - |A \text{ and } B|$$

we have, from the information given,

$$|A \text{ and } C| = |A| + |C| - |A \text{ or } C|$$
$$\geqslant 70 + 75 - 100 = 45,$$
$$|G \text{ and } T| = |G| + |T| - |G \text{ or } T|$$
$$\geqslant 80 + 85 - 100 = 65.$$

Consequently we have that

$$|A \text{ and } C \text{ and } G \text{ and } T|$$
$$= |A \text{ and } C| + |G \text{ and } T| - |(A \text{ and } C) \text{ or } (G \text{ and } T)|$$
$$\geqslant 45 + 65 - 100$$
$$= 10.$$

A Venn diagram illustrating this minimum percentage of 10 is shown in Fig. S1.12.

1.26    For each subset $A$ of $E$ let $|A|$ denote the number of elements in $A$. We make use of the formula

$$|A \cup B| = |A| + |B| - |A \cap B|.$$

**Fig.S1.12**

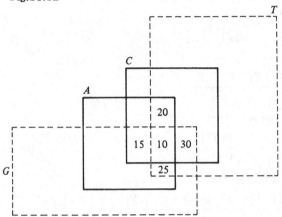

Note that, from this formula,

$$|A \cup B| = |A| + |B| \Leftrightarrow A \cap B = \emptyset.$$

Now let $A = X \cap Y$, $B = X' \cap Y$, $C = X \cap Y'$ and $D = X' \cap Y'$. Since $A \cap B = \emptyset$ and $A \cup B = Y$ we have, from the above, $|Y| = p + q$. Similarly, since $C \cap D = \emptyset$ and $C \cup D = Y'$ we have that $|Y'| = r + s$. Again, using the fact that $Y \cap Y' = \emptyset$ and $Y \cup Y' = E$, we deduce that $n = p + q + r + s$.

In order to relate this to the problem, consider Fig. S1.13 in which $X$ denotes the set of arts students and $Y$ denotes the set of boys.

**Fig.S1.13**

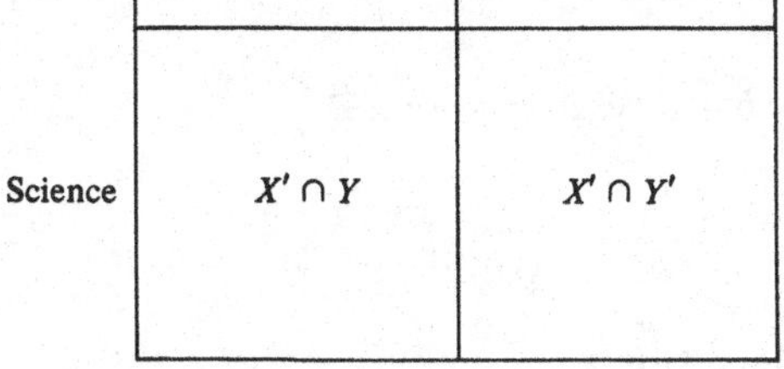

|  | Boys | Girls |
|---|---|---|
| Arts | $X \cap Y$ | $X \cap Y'$ |
| Science | $X' \cap Y$ | $X' \cap Y'$ |

Let $|X| = x$ and $|Y| = y$. What is given is that

$$\frac{|X \cap Y'|}{|Y'|} > \frac{|X \cap Y|}{|Y|},$$

so that we have

$$\frac{x - |X \cap Y|}{n - y} > \frac{|X \cap Y|}{y}.$$

What we have to prove is that

$$\frac{|X \cap Y'|}{|X|} > \frac{|X' \cap Y'|}{|X'|};$$

i.e., we have to show that

$$\frac{x - |X \cap Y|}{x} > \frac{n - x - y + |X \cap Y|}{n - x}.$$

This is clearly equivalent to

$$1 - \frac{|X \cap Y|}{x} > 1 - \frac{y - |X \cap Y|}{n - x}$$

which is the same as

$$\frac{|X \cap Y|}{x} < \frac{y - |X \cap Y|}{n - x}$$

which reduces to

$$n|X \cap Y| < xy.$$

Returning now to what we are given, we see that

$$xy - y|X \cap Y| > n|X \cap Y| - y|X \cap Y|,$$

which also reduces to $xy > n|X \cap Y|$. Hence the conclusion holds.

1.27   Let $x \in A$. Then we have

$$\begin{aligned} x = 2y \quad &\text{for some } y \in \mathbb{Z} \\ = 12y - 10y \\ = 6b + 10c \quad &\text{where } b = 2y \in \mathbb{Z}, c = -y \in \mathbb{Z} \end{aligned}$$

and so $x \in B$.

   Conversely, if $x \in B$ then

$$\begin{aligned} x = 6b + 10c \\ = 2y \quad &\text{where } y = 3b + 5c \in \mathbb{Z} \end{aligned}$$

and so $x \in A$.

*1.28*  (a)  $S = \mathbb{R}$ and $S = \mathbb{C}$;

(b)  $S = \mathbb{N}$ and $S = \mathbb{Z}$;

(c)  None; for $\emptyset \notin \mathbb{N}$, $\emptyset \notin \mathbb{Z}$, etc.

*1.29*  Yes; in fact $n_1\mathbb{Z} \cap n_2\mathbb{Z} = m\mathbb{Z}$ where $m = \mathrm{lcm}(n_1, n_2)$. To see this, let $t \in n_1\mathbb{Z} \cap n_2\mathbb{Z}$. Then $t$ is a common multiple of $n_1$ and $n_2$, whence it is a multiple of $m$ and hence belongs to $m\mathbb{Z}$. Thus $n_1\mathbb{Z} \cap n_2\mathbb{Z} \subseteq m\mathbb{Z}$. The reverse inclusion is immediate from the fact that every multiple of $m$ is a multiple of both $n_1$ and $n_2$.

In contrast, we have, for example, $2\mathbb{Z} \cup 3\mathbb{Z} \neq m\mathbb{Z}$ for any $m$. To see that this is so, suppose that we had $2\mathbb{Z} \cup 3\mathbb{Z} = m\mathbb{Z}$. Then from $2 \in m\mathbb{Z}$ and $3 \in m\mathbb{Z}$ we would have $2 = pm$, $3 = qm$ whence $1 = 3 - 2 = (q-p)m \in m\mathbb{Z}$. It follows from this that we must have $m = 1$; but clearly $1 \notin 2\mathbb{Z} \cup 3\mathbb{Z}$.

*1.30*  (a)  Given any $x \in A$ and any $B \subseteq A$ we have either $x \in B$ or $x \notin B$. Since there are these two possibilities for each of the $n$ elements of $A$ it follows that there are $2^n$ subsets of $A$.

(b)  Suppose that $n$ is even. There are $n/2$ odd integers in $A$ and hence $2^{n/2} - 1$ subsets with only odd integers. (Note that $\emptyset$ is a subset of this set of odd integers and must be excluded.) Hence $(2^n - 1) - (2^{n/2} - 1) = 2^n - 2^{n/2}$ subsets contain at least one even integer. Suppose now that $n$ is odd. There are $(n + 1)/2$ odd integers in $A$ and hence $2^n - 2^{(n+1)/2}$ subsets contain at least one even integer.

(c)  Suppose that $n$ is even. We have observed above that there are $2^{n/2} - 1$ subsets with only odd integers. Now there are $n/2$ even integers in $A$. Thus there are $(n/2)(2^{n/2} - 1)$ subsets consisting of odd integers and a single even integer. Add to this the $n/2$ singletons of even integers and we have a total of $(n/2)2^{n/2}$ subsets containing exactly one even integer. If now $n$ is odd then a similar argument applies : there are $2^{(n+1)/2} - 1$ subsets with only odd integers, and $(n - 1)/2$ even integers in $A$. Hence there are $2^{(n+1)/2}(n - 1)/2$ subsets with exactly one even integer.

*1.31*  One is looking for the longest chain of subsets

$$A_1 \subset A_2 \subset A_3 \subset \ldots$$

in $P(A)$. Take $A_1 = \emptyset$. $A_2$ must consist of a single element, $A_3$ must consist of two elements, etc. Thus we see that $k = n + 1$.

$$\sum_{i=1}^{k} |A_i| = \sum_{i=1}^{n+1} (i-1) = \sum_{i=1}^{n} i = \frac{n(n+1)}{2}.$$

1.32    One must treat separately the cases where $(x+1)(x-4)>0$ ar $(x+1)(x-4)<0$. The given set can be expressed as $A \cup B$ where

$$A = \{x \in \mathbb{R} \mid (x+1)(x-4)>0\}$$
$$\cap \{x \in \mathbb{R} \mid (x+1)(x-4)>-4\},$$
$$B = \{x \in \mathbb{R} \mid (x+1)(x-4)<0\}$$
$$\cap \{x \in \mathbb{R} \mid (x+1)(x-4)<-4\}.$$

Now

$$A = \{x \in \mathbb{R} \mid (x+1)(x-4)>0\}$$
$$= \{x \in \mathbb{R} \mid x<-1\} \cup \{x \in \mathbb{R} \mid x>4\}$$
$$= \,]-\infty,-1[\,\cup\,]4,\infty[,$$

and similarly

$$B = \{x \in \mathbb{R} \mid (x+1)(x-4)<-4\}$$
$$= \{x \in \mathbb{R} \mid x^2 - 3x < 0\}$$
$$= \{x \in \mathbb{R} \mid x(x-3)<0\}$$
$$= \,]0,3[.$$

Thus the required union of intervals is

$$]-\infty,-1[\,\cup\,]0,3[\,\cup\,]4,\infty[.$$

1.33    We have $(x-1)^2 = k(x+1)(x+3)$ if and only if

$$(k-1)x^2 + 2(2k+1)x + 3k - 1 = 0.$$

There is no real solution to this quadratic in $x$ if and only if

$$4(2k+1)^2 - 4(k-1)(3k-1)<0,$$

i.e., if and only if $k^2 + 8k < 0$, which is the case if and only if $k \in \,]-8,0[.$

# Solutions to Chapter 2

*2.1*  We have that

$$x^2 = x|y + 1| \Leftrightarrow x(x - |y + 1|) = 0$$
$$\Leftrightarrow x = 0 \text{ or } x = |y + 1|$$

whence we deduce that the graph of $S$ is as shown in Fig. S2.1.

**Fig.S2.1**

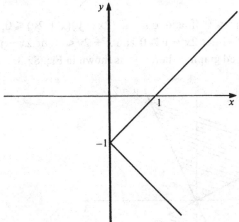

*2.2*  (a) Since

$$|x + y| \leqslant 1 \Leftrightarrow -1 \leqslant x + y \leqslant 1$$
$$\Leftrightarrow -1 - x \leqslant y \leqslant 1 - x$$

we see that the graph is as shown in Fig. S2.2.

**Fig.S2.2**

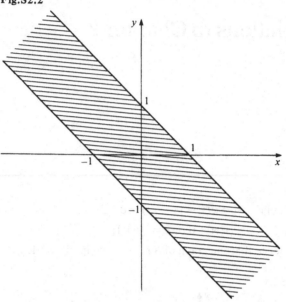

(*b*) $2x^2 + 3xy - 2y^2 \leqslant 0$ if and only if $(2x - y)(x + 2y) \leqslant 0$, which is the case if and only if either $2x - y \geqslant 0$ and $x + 2y \leqslant 0$, or $2x - y \leqslant 0$ and $x + 2y \geqslant 0$. The required graph is therefore as shown in Fig. S2.3.

**Fig.S2.3**

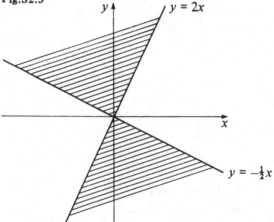

(*c*) Consider the expression

$$E \equiv (x - y)(\tfrac{1}{2}x - y)(\tfrac{1}{3}x - y).$$

Suppose first that $x \geq 0$. Then we have $E \geq 0$ if and only if $y \leq \frac{1}{3}x$ or $\frac{1}{2}x \leq y \leq x$. Suppose now that $x \leq 0$, say $x = -z$ where $z \geq 0$. We have

$$E \equiv -(z + y)(\tfrac{1}{2}z + y)(\tfrac{1}{3}z + y)$$

from which we see that $E \geq 0$ if and only if $y \leq -z = x$ or $\frac{1}{2}x = -\frac{1}{2}z \leq y \leq -\frac{1}{3}z = \frac{1}{3}x$. Thus the graph of the given relation is as shown in Fig. S2.4.

**Fig.S2.4**

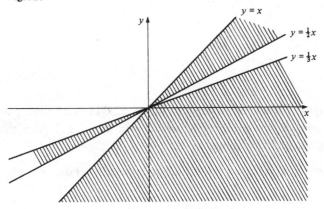

(*d*) The various inequalities are satisfied by the points $(x, y)$ contained in the region indicated in Fig. S2.5.

**Fig.S2.5**

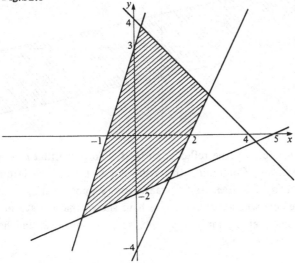

**Fig.S2.6**

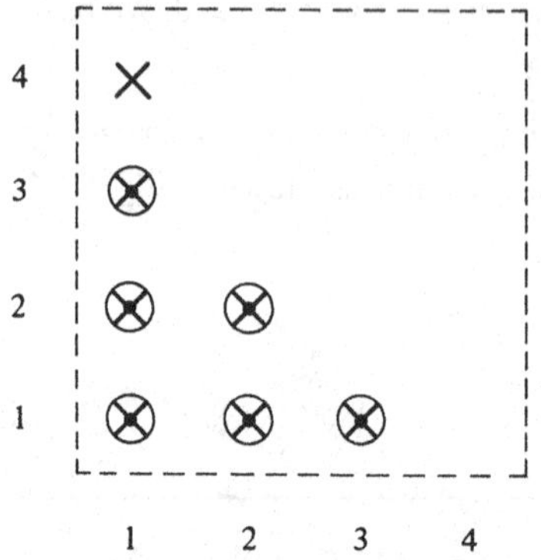

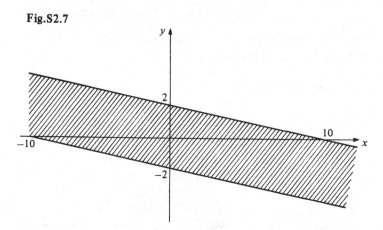

*2.3*   The graphs of $R$ and $S$ are depicted in Fig. S2.6 in which those points $(a, b)$ that belong to the graph of $R$ are denoted by $\odot$ and those that belong to the graph of $S$ are denoted by $\times$.

*2.4*   The graph of $R_1$ is shown in Fig. S2.7 and that of $R_2$ is shown in Fig. S2.8.

**Fig.S2.7**

*2.5*   Property $(a)$ is clearly the reflexive property. To prove that $\rho$ is symmetric, suppose that $a\rho b$. Combining this with the fact that $b\rho b$ (from $(a)$), we obtain from $(b)$ that $b\rho a$. As for transitivity, suppose that $a\rho b$ and $b\rho c$. By $(b)$ we have $c\rho a$, whence $a\rho c$ since we have shown that $\rho$ is symmetric. Thus $a\rho b$ and $b\rho c$ together imply that $a\rho c$, so $\rho$ is transitive, and hence is an equivalence relation on $A$.

**Fig.S2.8**

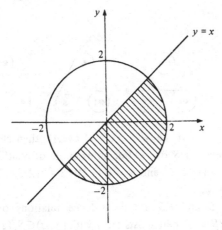

Every equivalence $R$ relation on $A$ clearly satisfies $(a)$. The condition $(b)$ is also satisfied; for if $aRb$ and $bRc$ then by transitivity $aRc$, and by symmetry $cRa$ follows.

2.6      $(a)$ To make $R$ reflexive on $X$ we must adjoint the elements $(b, b)$ and $(d, d)$.

     $(b)$ To make $R$ symmetric we must adjoin $(b, a), (c, a), (d, b)$.

     $(c)$ Note from the definition of $R$ that we have $aRb$, $aRc$ and $aRd$. Thus if $R$ is an equivalence relation on $X$ it must comprise all the elements of $X \times X$. So in order to make $R$ an equivalence relation on $X$ we must adjoin the remaining 11 elements of $X \times X$ to $R$.

Consider now the relation $S = \{(a, b), (a, c), (a, a), (c, c)\}$.

     $(a)$ To make $S$ reflexive on $X$ we must adjoin the elements $(b, b)$ and $(d, d)$.

     $(b)$ To make $S$ symmetric we must adjoin the elements $(b, a)$ and $(c, a)$.

     $(c)$ Note from the definition of $S$ that we have $aSb$ and $aSc$. The minimum number of elements that must be added to $S$ in order to make $S$ an equivalence relation is readily seen to be 6, namely the elements $(b, b), (b, a), (c, a), (b, c), (c, b)$ and $(d, d)$, the $S$-classes then being $\{a, b, c\}$ and $\{d\}$.

2.7   The answer is five. To see this, consider the number of equivalence classes that are possible. There is clearly only one equivalence relation with a single class; there are three distinct equivalence relations with a partition consisting of two classes; and only one with a partition consisting of three classes. These are depicted in Fig. S2.9.

**Fig.S2.9**

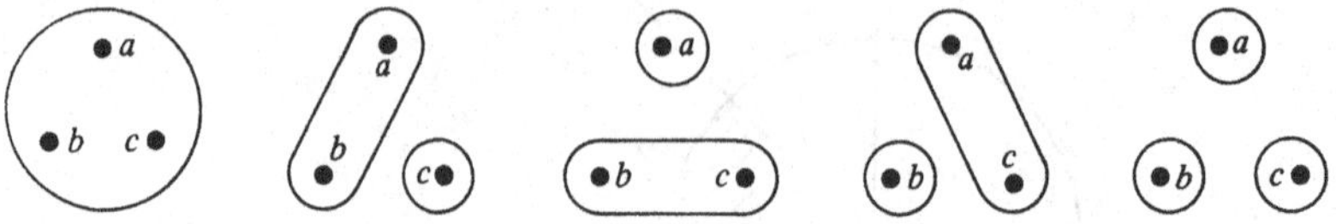

2.8    If $R$ is the identity relation on $A$, i.e. $R = \{(x, x) \mid x \in A\}$, then clearly we have $RS = SR$ for every relation $S$ on $A$. For an example of relations $R, S$ with $RS \neq SR$ consider $A = \{1, 2, 3\}$ and $R = \{(1, 2)\}$, $S = \{(2, 3)\}$. We have that $SR = \{(1, 3)\}$ and $RS = \emptyset$.

Suppose now that $R, S$ and $RS$ are equivalence relations on $A$. If $(x, y) \in RS$ then $(y, x) \in RS$ and there exists $z \in A$ with $(y, z) \in S, (z, x) \in R$. Now $(z, y) \in S$ and $(x, z) \in R$, and so $(x, y) \in SR$. This shows that $RS \subseteq SR$. A similar argument gives $SR \subseteq RS$ and equality follows.

Conversely, suppose that $R, S$ are equivalence relations with $RS = SR$. Since $(x, x) \in R$ and $(x, x) \in S$ for every $x \in A$ we see that $(x, x) \in RS$, so $RS$ is reflexive. To show that it is symmetric, let $(x, y) \in RS$. Then $(x, y) \in SR$ (since by hypothesis $RS = SR$) and so there exists $z \in A$ with $(x, z) \in R$, $(z, y) \in S$. Now $(z, x) \in R$, $(y, z) \in S$ since $R, S$ are symmetric, so $(y, x) \in RS$ as required. As for transitivity, let $(x, y) \in SR$, $(y, z) \in SR = RS$. Then for some $t, u \in A$ we have

$$(x, t) \in R, (t, y) \in S, (y, u) \in S, (u, z) \in R.$$

Since $S$ is transitive we deduce that $(t, u) \in S$ and hence

$$(x, u) \in SR = RS, (u, z) \in R.$$

The former gives $(x, w) \in S$, $(w, u) \in R$ for some $w \in A$. The transitivity of $R$ now yields

$$(x, w) \in S, (w, z) \in R,$$

from which it follows that $(x, z) \in RS = SR$ as required.

Finally, it is clear from the above that $RS$ is an equivalence relation if and only if $RS = SR$ if and only if $SR$ is an equivalence relation.

2.9    (a) Suppose that $R_1 \subseteq R_2$ and let $(x, y) \in SR_1$. For some $t \in X$ we have $(x, t) \in R_1$ and $(t, y) \in S$. Since then $(x, t) \in R_2$ we deduce that $(x, y) \in SR_2$ and hence that $SR_1 \subseteq SR_2$. Similarly, if $(x, y) \in R_1S$ then $(x, t) \in S$,

$(t, y) \in R_1 \subseteq R_2$ whence $(x, y) \in R_2 S$ and consequently $R_1 S \subseteq R_2 S$.

(b) Since $R_1 \subseteq R_1 \cup R_2$ we deduce from (a) that $SR_1 \subseteq S(R_1 \cup R_2)$. Similarly $SR_2 \subseteq S(R_1 \cup R_2)$ and hence $SR_1 \cup SR_2 \subseteq S(R_1 \cup R_2)$. To obtain the reverse inclusion, let $(x, y) \in SR_1 \cup SR_2$. Then either $(x, y) \in SR_1$ or $(x, y) \in SR_2$; i.e. either there exists $t \in X$ such that $(x, t) \in R_1, (t, y) \in S$ or there exists $t \in X$ such that $(x, t) \in R_2, (t, y) \in S$. Since $R_1, R_2 \subseteq R_1 \cup R_2$ it follows that in either event we have $(x, y) \in S(R_1 \cup R_2)$.

2.10      (a) For all $x \in X$ we have $(x, x) \in R_1$ and $(x, x) \in R_2$ and so $(x, x) \in R_1 \cap R_2$ whence $R_1 \cap R_2$ is reflexive on $X$. Suppose now that $(x, y) \in R_1 \cap R_2$. From $(x, y) \in R_1$ we have $(y, x) \in R_1$; and from $(x, y) \in R_2$ we have $(y, x) \in R_2$. Thus $(y, x) \in R_1 \cap R_2$ and so $R_1 \cap R_2$ is symmetric. As for transitivity, suppose that $(x, y), (y, z) \in R_1 \cap R_2$. From $(x, y), (y, z) \in R_1$ we deduce that $(x, z) \in R_1$, and from $(x, y), (y, z) \in R_2$ we deduce that $(x, z) \in R_2$. Thus $(x, z) \in R_1 \cap R_2$ and so $R_1 \cap R_2$ is transitive.

(b) Take, for example, $X = \{1, 2, 3\}$ and let

$$R_1 = \{(1, 1), (2, 2), (3, 3), (1, 2), (2, 1)\},$$
$$R_2 = \{(1, 2), (2, 2), (3, 3), (1, 3), (3, 1)\}.$$

Then clearly $R_1$ and $R_2$ are equivalence relations on $X$. However, $R_1 \cup R_2$ is not an equivalence relation since we have, for example, $(3, 1), (1, 2) \in R_1 \cup R_2$ but $(3, 2) \notin R_1 \cup R_2$.

For the last part, consider $X = \{1, 2\}$ and let

$$R_1 = \{(1, 1), (2, 2), (1, 2), (2, 1)\},$$
$$R_2 = \{(1, 1), (2, 2)\}.$$

Clearly, $R_1$ and $R_2$ are equivalence relations on $X$ with $R_1 \neq R_2$. Here we have $R_1 \cup R_2 = R_1$ which is an equivalence relation.

2.11      That $\alpha$ is an equivalence relation on $\mathbb{N}$ is immediate from the fact that the relation mod 7 is an equivalence relation. Now we have

$$a \alpha b \Leftrightarrow (a - b)(a + b) \equiv 0 \pmod 7$$
$$\Leftrightarrow a \equiv b \pmod 7 \text{ or } a \equiv -b \pmod 7.$$

Since, modulo 7, we have $6 \equiv -1$, $5 \equiv -2$, $4 \equiv -3$ it follows that there are four $\alpha$-classes, namely $[0], [1], [2], [3]$.

2.12      It is immediate from the definition that $\rho$ is reflexive and symmetric on $S \times S$. To see that it is also transitive, we note that if $(a, b) \rho (c, d)$ and $(c, d) \rho (e, f)$ then $c^2 b = a^2 d$ and $e^2 d = c^2 f$. Since by definition $a, c, e \neq 0$

we deduce that

$$\frac{b}{a^2} = \frac{d}{c^2} = \frac{f}{e^2},$$

whence $e^2 b = a^2 f$ and hence $(a, b)\rho(e, f)$. Thus $\rho$ is an equivalence relation.

Now, since

$$(x, y)\rho(c, d) \Leftrightarrow c^2 y = x^2 d$$

$$\Leftrightarrow y = \frac{d}{c^2} x^2,$$

we see that the $\rho$-classes can be described geometrically as parabolae passing through $(0, 0)$ with $(0, 0)$ deleted (since $0 \notin S$ by definition).

The proof that $\sigma$ is also an equivalence relation on $S \times S$ is entirely similar to the above. In this case, we have

$$(x, y)\sigma(c, d) \Leftrightarrow c^4 y^2 = x^4 d^2$$

$$\Leftrightarrow y^2 = \frac{d^2}{c^4} x^4$$

$$\Leftrightarrow \left(y - \frac{d}{c^2} x^2\right)\left(y + \frac{d}{c^2} x^2\right) = 0$$

$$\Leftrightarrow y = \frac{d}{c^2} x^2 \text{ or } y = -\frac{d}{c^2} x^2.$$

Thus we see that each $\sigma$-class consists of a pair of parabolae, one of these being above the $x$-axis and the other consisting of its reflection in the $x$-axis (with, as before, the point $(0, 0)$ deleted). Hence each $\sigma$-class consists of two $\rho$-classes.

Finally, $\tau$ is not an equivalence relation on $\mathbb{R} \times \mathbb{R}$ since, for example, we have $(1, 1)\tau(0, 0)$ and $(0, 0)\tau(0, 5)$, but $(1, 1), (0, 5)$ are not $\tau$-related.

2.13   That $\sim$ is reflexive and symmetric on $\mathbb{C} \setminus \{0\}$ is immediate from the definition. To show that $\sim$ is transitive, let $z_1 \sim z_2$ and $z_2 \sim z_3$. Then we have

$$\frac{|z_1|^2 + 1}{|z_1|} = \frac{|z_2|^2 + 1}{|z_2|} = \frac{|z_3|^2 + 1}{|z_3|},$$

from which we see that $z_1 \sim z_3$.

If now $a \in \mathbb{R}$ is such that $0 < a < 1$ we have, writing $z = x + iy$, that

$$z \sim a \Leftrightarrow |z|(a^2 + 1) = a(|z|^2 + 1)$$

$$\Leftrightarrow (x^2 + y^2)(a^2 + 1)^2 = a^2(x^2 + y^2 + 1)^2$$

$$\Leftrightarrow (x^2 + y^2 - a^2)(a^2 x^2 + a^2 y^2 - 1) = 0$$

$$\Leftrightarrow x^2 + y^2 = a^2 \text{ or } x^2 + y^2 = \frac{1}{a^2}.$$

Thus we see that the $\sim$-class of $a \in {]}0, 1{[}$ consists of two circles in the Argand diagram (Fig. S2.10). These are concentric at the origin, one passes through $a$ and the other through $1/a$.

**Fig.S2.10**

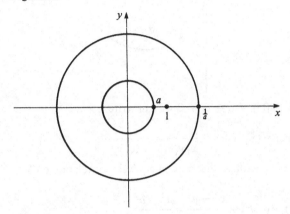

2.14   Observe first that for $z \in \mathbb{C} \setminus \{0\}$ we have $z\bar{z} = |z|^2 \neq 0$. Consequently we can write

$$z_1 \sim z_2 \Leftrightarrow \frac{z_1 + \bar{z}_1}{z_1 \bar{z}_1} = \frac{z_2 + \bar{z}_2}{z_2 \bar{z}_2},$$

from which it is easy to see that $\sim$ is an equivalence relation.

If now $a$ is a non-zero number on the real axis we have, writing $z = x + iy$,

$$z \sim a \Leftrightarrow \frac{2x}{x^2 + y^2} = \frac{2a}{a^2} = \frac{2}{a}$$

$$\Leftrightarrow x^2 + y^2 = ax$$

$$\Leftrightarrow (x - \tfrac{1}{2}a)^2 + y^2 = \tfrac{1}{4}a^2.$$

Thus the $\sim$-class of $a$ is a circle with centre at the point $(\tfrac{1}{2}a, 0)$ and of radius $\tfrac{1}{2}a$ (Fig. S2.11).

**Fig.S2.11**

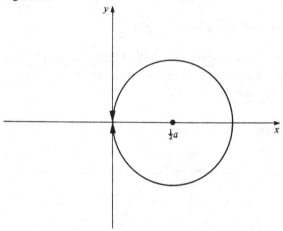

2.15        (*a*) Since $x \neq 0, y \neq 0 \Rightarrow xy \neq 0$ we can write

$$(x_1, y_1) \sim (x_2, y_2) \Leftrightarrow \frac{x_1^2 - y_1^2}{x_1 y_1} = \frac{x_2^2 - y_2^2}{x_2 y_2},$$

from which it is immediate that $\sim$ is an equivalence relation on *S*.

(*b*)

$$(x, y) \sim (a, b) \Leftrightarrow \frac{x^2 - y^2}{xy} = \frac{a^2 - b^2}{ab}$$

$$\Leftrightarrow abx^2 - aby^2 = xya^2 - xyb^2$$

$$\Leftrightarrow (ay - bx)(ax + by) = 0$$

$$\Leftrightarrow ay - bx = 0 \quad \text{or} \quad ax + by = 0$$

$$\Leftrightarrow \frac{y}{x} = \frac{b}{a} \quad \text{or} \quad \frac{y}{x} = -\frac{a}{b}.$$

(*Note:* Another way of obtaining this result is to write $(x, y) \sim (a, b)$ as

$$\frac{x}{y} - \frac{y}{x} = \frac{a}{b} - \frac{b}{a}.$$

Under the definitions

$$t = \frac{y}{x} \quad \text{and} \quad \alpha = \frac{a}{b} - \frac{b}{a},$$

this can be transformed into the quadratic equation $t^2 + t\alpha - 1 = 0$ whose roots are $b/a$ and $-a/b$.)

46

(c) We have

$$(x,y) \sim (2,1) \Leftrightarrow \frac{y}{x} = \frac{1}{2} \quad \text{or} \quad \frac{y}{x} = -2,$$

so the $\sim$-class of $(2, 1)$ can be described geometrically as the perpendicular line-pair $y = \frac{1}{2}x$, $y = -2x$ with the origin deleted (since $(0, 0) \notin S$). See Fig. S2.12.

**Fig.S2.12**

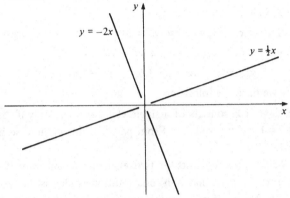

2.16     (a) Use calculus to sketch the graph of $f$ (Fig. S2.13). The minimum value of $x^2 + ax + a^2$ is attained when $0 = 2x + a$, i.e. when $x = -\frac{1}{2}a$. The minimum value is then $\frac{3}{4}a^2$.

**Fig.S2.13**

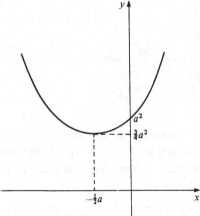

(i) The statement $A = \emptyset$ is equivalent to saying that the line $y = 1$ does not meet the graph of $f$. This is the case if and only if $\frac{3}{4}a^2 > 1$, i.e. $|a| > 2/\sqrt{3}$.

(ii) $|A| = 1$ is equivalent to saying that the line $y = 1$ meets the graph of $f$ exactly once. This is the case if and only if $\frac{3}{4}a^2 = 1$, i.e. $|a| = 2/\sqrt{3}$.

(iii) $|A| = 2$ is equivalent to saying that the line $y = 1$ meets the graph of $f$ in two points. This is the case if and only if $\frac{3}{4}a^2 < 1$, i.e. $|a| < 2/\sqrt{3}$.

(*b*) That $S$ is an equivalence relation on $\mathbb{R}$ is immediately seen from the fact that

$$x \equiv y(S) \Leftrightarrow x^3 - x = y^3 - y.$$

As for the $S$-class of $x \in \mathbb{R}$, we observe that

$$y \equiv x(S) \Leftrightarrow (x - y)(x^2 + xy + y^2 - 1) = 0$$
$$\Leftrightarrow y = x \quad \text{or} \quad x^2 + xy + y^2 - 1 = 0.$$

Consequently we have the following.

(i) The $S$-class of $x$ consists of a single element if and only if the equation $x^2 + xy + y^2 = 1$ has no solution. From part (*a*), this is the case if and only if $|x| > 2/\sqrt{3}$.

(ii) The $S$-class of $x$ consists of two elements if and only if either the equation $x^2 + xy + y^2 = 1$ has only one solution (which is the case precisely when $|x| = 2/\sqrt{3}$), or if the equations $y = x$ and $x^2 + xy + y^2 = 1$ have a solution in common (which is the case precisely when $|x| = 1/\sqrt{3}$).

(iii) The $S$-class of $x$ consists of three elements otherwise.

2.17     (*a*) We have that

$$xRy \Leftrightarrow x^2 + 2x = y^2 + 2y,$$

from which it is immediate that $R$ is an equivalence relation on $\mathbb{R}$. Since

$$xR0 \Leftrightarrow x(x + 2) = 0,$$

it is clear that the $R$-class of $0$ is $\{0, -2\}$. Likewise,

$$xR1 \Leftrightarrow x^2 + 2x = 3$$
$$\Leftrightarrow (x - 1)(x + 3) = 0,$$

so the $R$-class of $1$ is $\{1, -3\}$.

(*b*) The given argument breaks down at the point where cancellation by $x - y$ takes place. This step is valid only when $x - y \neq 0$. In fact the relation $S$ is far from being an equivalence relation since, for example, we have $xSx$ if and only if $x = -1$.

2.18    Suppose that $M$ contains no subset $\{b_1, b_2, \ldots, b_{n+1}\}$ of $n + 1$ elements with the property that $b_j S b_k$ for $j \neq k$. Let $B$ be the subset of $M$ consisting of the $n + 1$ smallest integers in $M$. There must be a smallest $b_{11} \in B$ such that $b_{11}$ divides some element of $B$. Replace $b_{11}$ in $B$ by the next smallest element of $M$ and place $b_{11}$ into a set $A_1$. Again there is a least element $b$ in $B$ which divides some other element of $B$. If $b_{11} \equiv b$ then denote $b$ by $b_{12}$ and put it into $A_1$; if $b_{11} \not\equiv b$ then denote $b$ by $b_{21}$ and put it into a set $A_2$. Now add to $B$ the next smallest element of $M$ and continue the process.

Note that at most $n$ sets $A_1, A_2, \ldots, A_n$ can be built up in this way since each time we remove an element $b$ from $B$ we leave an element $\bar{b} \in B$ with $b \equiv \bar{b}$ and at this stage $B$ contains $n$ elements.

Since $M$ contains $mn + 1$ elements, one of the subsets $A_i$ must contain at least $m + 1$ elements when the process is complete. If this subset is $\{a_1, a_2, \ldots, a_{m+1}\}$ then by its construction we have $a_i \equiv a_{i+1}$ for $1 \leqslant i \leqslant m$.

2.19  Let $U$, $V$ be distinct constituents. Then there is an $A_i$ with $A_i$ appearing in $U$ and $A_i'$ appearing in $V$. Then $U \subseteq A_i$ and $V \subseteq A_i'$ and so
$$U \cap V \subseteq A_i \cap A_i' = \emptyset,$$
whence the constituents are pairwise disjoint. Consider now any $x \in X$. Either $x \in A_1$ or $x \in A_1'$; and either $x \in A_2$ or $x \in A_2'$; and so on, so for each $i$ there is an $\epsilon_i \in \{0, 1\}$ with $x \in A_i^{\epsilon_i}$. Thus $x$ belongs to some constituent and the constituents partition $X$.

The Venn diagram for $A \setminus (B \setminus C)$ is as shown in Fig. S2.14.

**Fig.S2.14**

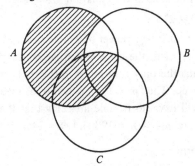

Expressed as a union of constituents with respect to $A, B, C$,
$$A \setminus (B \setminus C) = (A \cap B \cap C) \cup (A \cap B' \cap C) \cup (A \cap B' \cap C').$$
Also, $(A \cap B) \setminus C$ is the constituent $A \cap B \cap C'$. Hence, since $A \setminus (B \setminus C)$ is the union of constituents different from $A \cap B \cap C'$, $x$ and $y$ must lie in

different members of the partition. Then $x, y$ lie in different $\equiv$-classes and so cannot be equivalent.

2.20    By definition,
$$\rho = \{(a, b) \mid a - \sqrt{(a + 1)} \leqslant b - \tfrac{1}{4} \leqslant a + \sqrt{(a + 1)}\}.$$
We thus have that
$$(a, b) \in \rho \Leftrightarrow -\sqrt{(a + 1)} \leqslant b - a - \tfrac{1}{4} \leqslant \sqrt{(a + 1)}$$
$$\Leftrightarrow |b - a - \tfrac{1}{4}| \leqslant \sqrt{(a + 1)}$$
$$\Leftrightarrow (b - a - \tfrac{1}{4})^2 \leqslant a + 1$$
$$\Leftrightarrow a^2 + b^2 - 2ab - \tfrac{1}{2}a - \tfrac{1}{2}b - \tfrac{15}{16} \leqslant 0.$$
It is clear from this that $\rho$ is reflexive (since $a, b \geqslant 0$) and symmetric. It is not transitive, however, since for example $(0, 1), (1, 2) \in \rho$ but $(0, 2) \notin \rho$.

2.21    The number of integers in the interval $[n, n + m]$ is $m + 1$ and so, from the information given, for every integer $t \in [n, n + m]$ there is a member of the partition that has precisely $t$ elements. We thus have
$$n + (n + 1) + (n + 2) + \ldots + (n + m) = 1000.$$
Now the expression on the left is
$$(m + 1)n + \sum_{r=1}^{m} r = (m + 1)n + \tfrac{1}{2}m(m + 1)$$
$$= \tfrac{1}{2}(m + 1)(2n + m)$$
and so we have to find all integer solutions of the equation
$$(m + 1)(2n + m) = 2000 = 2^4 5^3.$$
Suppose that $m$ is odd. Then $m + 1$ must be a power of 2. The only possibility is $m + 1 = 2^4$ since, as is readily verified, the other potential values give the contradiction that $2n$ is odd. Thus the only solution in this case is $m = 15$, $n = 55$. Suppose now that $m$ is even. Then $m + 1$ is a power of 5. Listing the various possibilities, the reader will have no trouble in seeing that the other possible solutions to the problem are $m = 4, n = 198$ and $m = 24, n = 8$.

2.22    Since $\sim$ can be expressed in the form
$$(x, y) \sim (a, b) \Leftrightarrow \frac{x^2}{y^2} = \frac{a^2}{b^2} \Leftrightarrow \frac{x^2}{a^2} = \frac{y^2}{b^2}$$
it is clear that $\sim$ is an equivalence relation. Moreover, we have that $(x, y) \sim (a, b)$ if and only if

$$\left(\frac{x}{a} - \frac{y}{b}\right)\left(\frac{x}{a} + \frac{y}{b}\right) = 0,$$

which is the case if and only if

$$\frac{x}{a} = \frac{y}{b} \quad \text{or} \quad \frac{x}{a} = -\frac{y}{b}.$$

Writing $x/a = k$, these equations can be written in the form $x = ka, y = \pm kb$, which is of course the line-pair $y = \pm(b/a)x$.

The $\sim$-class of $(2, 1)$ is represented in Fig. S2.15 as a line-pair with the origin deleted.

**Fig.S2.15**

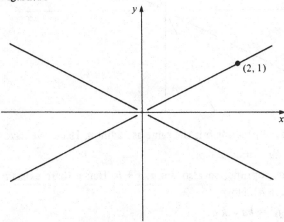

**2.23** From the definition of $E$ we see that

$$(x, y) \equiv (z, t) \Leftrightarrow \frac{x^2 + xy + y^2}{xy} = \frac{z^2 + zt + t^2}{zt},$$

whence $\equiv$ is clearly an equivalence relation on $E$.

If now $m \neq 0$ we have

$$(x, y) \equiv (1, m) \Leftrightarrow \frac{x^2 + xy + y^2}{xy} = \frac{1 + m + m^2}{m}$$

$$\Leftrightarrow mx^2 + my^2 = xy + xym^2$$

$$\Leftrightarrow (mx - y)(x - my) = 0$$

$$\Leftrightarrow y = mx \quad \text{or} \quad y = \frac{1}{m}x.$$

Thus the $\equiv$-class of $(1, m)$ consists of the line-pair $y = mx, y = x/m$ with the origin deleted (Fig. S2.16). The equivalence class containing a general point $(a, b)$ is the same as that containing the point $(1, m)$ where $m = b/a$.

**Fig.S2.16**

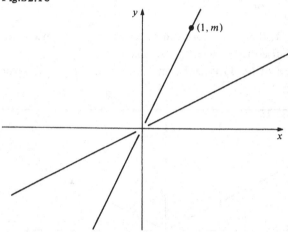

2.24    Modulo $a + b$, subtracting $b$ is the same as adding $a$. Hence we have that
$$f_k = ka \quad \text{mod } a + b.$$
Since $a, b$ are coprime, so also are $a, a + b$. Hence there exists $c$ such that $ac \equiv 1 \pmod{a + b}$. Now
$$f_k \equiv f_{k'} \Rightarrow ka \equiv k'a$$
$$\Rightarrow (k - k')a \equiv 0$$
$$\Rightarrow (k - k')ac \equiv 0$$
$$\Rightarrow k - k' \equiv 0.$$
Since we can assume that $1 \leqslant k \leqslant a + b$ and $1 \leqslant k' \leqslant a + b$, this gives $k = k'$ as required.

   If $a, b$ fail to be coprime then by definition each $f_k$ is a multiple of $d = \text{hcf}(a, b)$ and the set of $f_k$ does not form a complete set of representatives, since the class of 1 contains no integer divisible by $d$ and hence no $f_k$. In this case there exists $c$ such that $ac \equiv d \pmod{a + b}$ so that, modulo $a + b$, $d = f_c$. It follows that $pd = f_{pc} \pmod{a + b}$ and hence every multiple of $d$ is some $f_k \pmod{a + b}$.

2.25     (a)  See Fig. S2.17.

**Fig.S2.17**

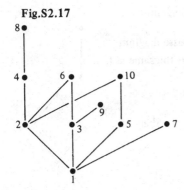

(*b*) See Fig. S2.18.

**Fig.S2.18**

2 ●
4 ●
6 ●
8 ●
10 ●    ● 1
9 ●
7 ●
5 ●
3 ●

(*c*) See Fig. S2.19.

**Fig.S2.19**

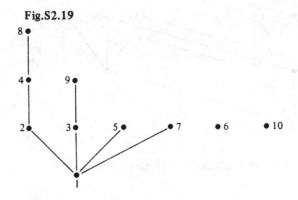

*2.26*      (*a*)  $|E| = 0$ so $|\mathbf{P}(E)| = 1$. Hasse diagram : ●

(*b*)  $|E| = 1$ so $|\mathbf{P}(E)| = 2^1 = 2$. Hasse diagram :
(*c*)  $|E| = 1$ so the Hasse diagram is the same as (*b*).

(*d*)  $|E| = 2$ so $|\mathbf{P}(E)| = 2^2 = 4$. Hasse diagram :

(*e*)  $|E| = 3$ so $|\mathbf{P}(E)| = 2^3 = 8$. Hasse diagram :

*2.27*   The set of prime factors of 210 is {2, 3, 5, 7}, and there are 16 positive divisors, as shown in the Hasse diagram (Fig. S2.20).

**Fig.S2.20**

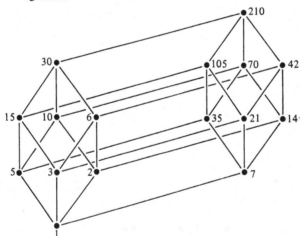

*2.28*   Consider the elements of $E$ arranged in a rectangular array

$$x_{11} \quad x_{12} \quad \ldots \quad x_{1n}$$
$$x_{21} \quad x_{22} \quad \ldots \quad x_{2n}$$
$$\vdots \qquad \vdots \qquad \quad \vdots$$
$$x_{m1} \quad x_{m2} \quad \ldots \quad x_{mn}$$

By definition, $y_i$ is the greatest element in the $i$th row; and $z_j$ is the least element in the $j$th column. Thus we have

$$(\forall i, j) \; y_i \geqslant x_{ij} \geqslant z_j.$$

It follows that, for all $j$, min $\{ y_i \mid 1 \leqslant i \leqslant m \geqslant z_j$ and hence that

$$\min \{ y_i \mid 1 \leqslant i \leqslant m \} \geqslant \max \{ z_j \mid 1 \leqslant j \leqslant n \}.$$

Applying the above argument to the problem of the soldiers, it is clear that Sergeant Mintall is taller than Corporal Max Small.

2.29   It is immediate from $x \leqslant x$ that $\lfloor x \rfloor R \lfloor x \rfloor$, so that $R$ is reflexive. Suppose now that $[x]R[y]$ and $[y]R[x]$. Then for every $a \in [x]$ there exists $b \in [y]$ with $a \leqslant b$; and for every $b' \in [y]$ there exists $a' \in [x]$ such that $b' \leqslant a'$. Taking $b' = b$ and using the given property, we obtain $a \equiv b$ and hence $[x] = [y]$. Thus $R$ is anti-symmetric. Finally, if $[x]R[y]$ and $[y]R[z]$ then for every $a \in [x]$ there exists $b \in [y]$ with $a \leqslant b$, and for every $b \in [y]$ there exists $c \in [z]$ with $b \leqslant c$. Thus, for every $a \in [x]$ there exists $c \in [z]$ such that $a \leqslant c$, whence $[x]R[z]$ and $R$ is transitive. Thus $R$ is an order relation on $\bar{E}$.

# Solutions to Chapter 3

**3.1** (a) $P \circ C$; (b) $C \circ P$; (c) $S \circ C$; (d) $C \circ S$; (e) $C \circ C$; (f) $C \circ [P \cdot (P \circ C)]$; (g) $C \circ S \circ C$; (h) $S \circ C \circ P \circ C$; (i) $S \circ P \circ C$; (j) $(S \circ P \circ S \circ C \circ P) \cdot (P \circ S)$.

**3.2** We have, for example,

$$(f_6 \circ f_5)(x) = f_6 \left( \frac{1}{1-x} \right) = \frac{\dfrac{1}{1-x}}{\dfrac{1}{1-x} - 1}$$

$$= \frac{1}{1 - (1-x)}$$

$$= \frac{1}{x},$$

and hence $f_6 \circ f_5 = f_4$. Proceed similarly with all other composite pairs to show that, for all $i$ and $j$, $f_i \circ f_j$ belongs to the set of six given mappings. Check your answers with the following 'composition table' of which the interpretation is that $f_i \circ f_j$ appears at the intersection of the row headed by $f_i$ and the column headed by $f_j$:

| $\circ$ | $f_1$ | $f_2$ | $f_3$ | $f_4$ | $f_5$ | $f_6$ |
|---|---|---|---|---|---|---|
| $f_1$ | $f_1$ | $f_2$ | $f_3$ | $f_4$ | $f_5$ | $f_6$ |
| $f_2$ | $f_2$ | $f_1$ | $f_4$ | $f_3$ | $f_6$ | $f_5$ |
| $f_3$ | $f_3$ | $f_6$ | $f_5$ | $f_2$ | $f_1$ | $f_4$ |
| $f_4$ | $f_4$ | $f_5$ | $f_6$ | $f_1$ | $f_2$ | $f_3$ |
| $f_5$ | $f_5$ | $f_4$ | $f_1$ | $f_6$ | $f_3$ | $f_2$ |
| $f_6$ | $f_6$ | $f_3$ | $f_2$ | $f_5$ | $f_4$ | $f_1$ |

56

## Solutions to Chapter 3

3.3    Let $g_1, g_2, g_3, g_4 : \mathbb{R} \to \mathbb{R}$ be defined by

$$g_1(x) = 1 - x, \quad g_2(x) = x^3, \quad g_3(x) = 1 + x, \quad g_4(x) = x^{1/3}.$$

Then we have $f = g_4 \circ g_3 \circ g_2 \circ g_1$.

3.4    (a) $R$ is reflexive on $X$ if and only if $xRx$ for all $x \in X$, i.e. if and only if $x = f(x)$ for all $x \in X$, i.e. if and only if $f = \mathrm{id}_X$.

(b) The criterion for symmetry is

$$xRy \Rightarrow yRx$$

which is equivalent to

$$y = f(x) \Rightarrow x = f(y).$$

Suppose that $R$ is symmetric and let $x \in X$. If $f(x) = y$ then from the above we have $x = f(y)$. But then $f^2(x) = f(y) = x$ and consequently $f^2 = \mathrm{id}_X$. Conversely, if $f^2 = \mathrm{id}_X$ then from $y = f(x)$ we deduce that $f(y) = f^2(x) = x$ whence we see that $R$ is symmetric.

(c) The criterion for transitivity is

$$(xRy \text{ and } yRz) \Rightarrow xRz$$

which is equivalent to

$$(y = f(x) \text{ and } z = f(y)) \Rightarrow z = f(x).$$

Suppose that $R$ is transitive and let $x \in X$. Let $f(x) = y$ and let $f(y) = z$. By the above we have $z = f(x)$. But $z = f(y) = f^2(x)$. Hence we see that $f = f^2$. Conversely, if $f = f^2$ then the conditions $y = f(x)$ and $z = f(y)$ give $z = f^2(x) = f(x)$ and hence $R$ is transitive.

3.5    We are given that

$$f = \{(x, y) \in X \times X \mid y = f(x)\},$$
$$g = \{(y, x) \in X \times X \mid y = f(x)\}.$$

Now $(a, b) \in g \circ f$ if and only if there exists $t \in X$ with $(a, t) \in f$ and $(t, b) \in g$. For this to be so, it is necessary and sufficient that $f(a) = f(b)(=t)$. Thus we have

$$(a, b) \in g \circ f \Leftrightarrow f(a) = f(b).$$

Now since clearly $f(x) = f(x)$ we have $(x, x) \in g \circ f$ and so $g \circ f$ is reflexive. Also, if $(x, y) \in g \circ f$ then $f(x) = f(y)$ whence $f(y) = f(x)$ and $(y, x) \in g \circ f$, so that $g \circ f$ is symmetric. Finally, if $(x, y), (y, z) \in g \circ f$ then $f(x) = f(y)$, $f(y) = f(z)$ so $f(x) = f(z)$ and $(x, z) \in g \circ f$, whence $g \circ f$ is transitive.

3.6    The given relation is clearly reflexive since $f(x) = f(x)$ holds trivially for all $x \in A$. It is also symmetric since $f(x) = f(y)$ gives $f(y) = f(x)$. That it is

57

transitive is equally immediate : $f(x) = f(y)$ and $f(y) = f(z)$ imply that $f(x) = f(z)$.

Now $(x, y) R_f (1, 0)$ if and only if $f(x, y) = f(1, 0) = (1, 0)$, i.e.

$$\frac{x}{\sqrt{(x^2 + y^2)}} = 1 \quad \text{and} \quad \frac{y}{\sqrt{(x^2 + y^2)}} = 0.$$

It follows that the $R_f$-class of $(1, 0)$ is

$$\{(x, 0) \mid x > 0\}.$$

For a general $(a, b) \neq (0, 0)$ we have that $(x, y)$ belongs to the $R_f$-class of $(a, b)$ if and only if

$$\frac{x}{\sqrt{(x^2 + y^2)}} = \frac{a}{\sqrt{(a^2 + b^2)}} \quad \text{and} \quad \frac{y}{\sqrt{(x^2 + y^2)}} = \frac{b}{\sqrt{(a^2 + b^2)}}.$$

These conditions give $y/x = b/a$ where $x$, $a$ have the same sign and $y$, $b$ have the same sign. This last equation may be written $y = (b/a)x$. We conclude that the $R_f$-class of $(a, b)$ is the half-line from (but excluding) the origin which passes through $(a, b)$. For $(a, b) = (0, 0)$ the $R_f$-class is the singleton $\{(0, 0)\}$.

3.7  (a)  $x^2 \leqslant y$

    (i) See Fig. S3.1.

    (ii) The domain is $\mathbb{R}$.

        The image is $\{y \in \mathbb{R} \mid y \geqslant 0\}$.

    (iii) It is not a mapping.

**Fig.S3.1**

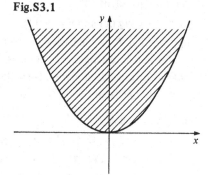

    (b)  $\sin x = y$

    (i) See Fig. S3.2.

    (ii) The domain is $\mathbb{R}$.

        The image is $[-1, 1]$.

    (iii) It is a mapping.

**Fig.S3.2**

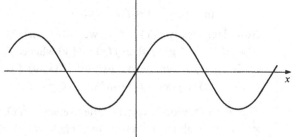

## Solutions to Chapter 3

(c)  $x = \sin y$

  (i) See Fig. S3.3.

  (ii) The domain is $[-1, 1]$.

     The image is $\mathbb{R}$.

  (iii) It is not a mapping.

**Fig.S3.3**

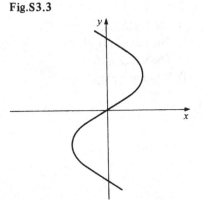

(d)  $x - 2 \leqslant y \leqslant x + 1$

  (i) See Fig. S3.4.

  (ii) The domain is $\mathbb{R}$.

     The image is $\mathbb{R}$.

  (iii) It is not a mapping.

**Fig.S3.4**

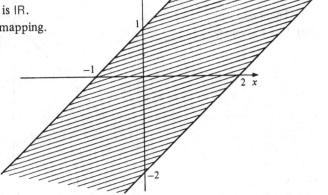

(e)  $y = |x|$

  (i) See Fig. S3.5.

  (ii) The domain is $\mathbb{R}$.

     The image is $\{y \in \mathbb{R} \mid y \geqslant 0\}$.

  (iii) It is a mapping.

**Fig.S3.5**

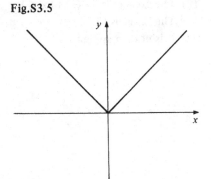

(*f*) $x + y = 1$

  (i) See Fig. S3.6.

  (ii) The domain is IR.

     The image is IR.

  (iii) It is a mapping.

**Fig.S3.6**

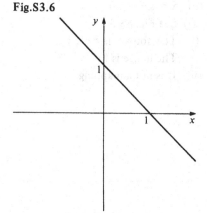

(*g*) $|x| + y = 1$

  (i) See Fig. S3.7.

  (ii) The domain is IR.

     The image is $\{y \in IR \mid y \leqslant 1\}$.

  (iii) It is a mapping.

**Fig.S3.7**

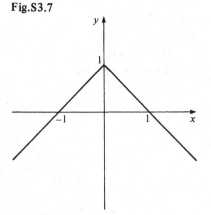

(*h*) $x + |y| = 1$

  (i) See Fig. S3.8.

  (ii) The domain is $\{x \in IR \mid x \leqslant 1\}$.

     The image is IR.

  (iii) It is not a mapping.

**Fig.S3.8**

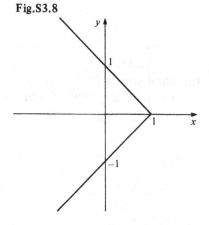

(*i*)  $|x| + |y| = 1$
  (i) See Fig. S3.9.
  (ii) The domain is $[-1, 1]$.
     The image is $[-1, 1]$.
  (iii) It is not a mapping.

**Fig.S3.9**

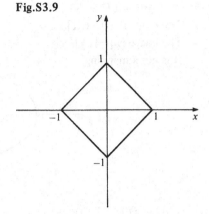

(*j*)  $|y| \leqslant |x| \leqslant 1$
  (i) See Fig. S3.10.
  (ii) The domain is $[-1, 1]$.
     The image is $[-1, 1]$.
  (iii) It is not a mapping.

**Fig.S3.10**

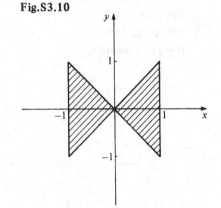

(*k*)  $y = |x| - [\![x]\!]$
  (i) See Fig. S3.11.
  (ii) The domain is ℝ.
     The image is the infinite union
     $[0, 1[ \cup ]1, 2] \cup ]3, 4]$
                 $\cup ]5, 6] \cup \ldots$
  (iii) It is a mapping.

**Fig.S3.11**

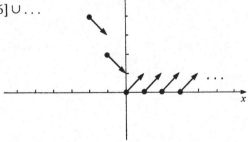

*3.8*    (*a*)   $x^2 + 4y^2 = 1$ (Fig. S3.12)

**Fig.S3.12**

The domain is $[-1, 1]$.

The image is $[-\frac{1}{2}, \frac{1}{2}]$.

It is not a mapping.

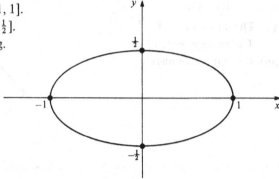

(*b*)   $x^2 = y^2$ (Fig. S3.13)

**Fig.S3.13**

The domain is $\mathbb{R}$.

The image is $\mathbb{R}$.

It is not a mapping.

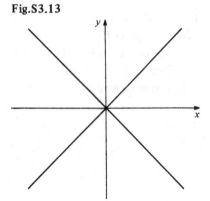

(*c*)   $y \geqslant 0, y \leqslant x, x + y \leqslant 1$
(Fig. S3.14)

**Fig.S3.14**

The domain is $[0, 1]$.

The image is $[0, \frac{1}{2}]$.

It is not a mapping.

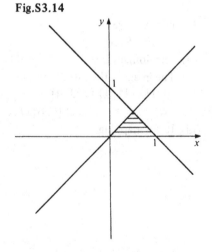

62

(*d*) $x^2 + y^2 \leqslant 1, x \geqslant 0$
(Fig. S3.15)
The domain is $[0, 1]$.
The image is $[-1, 1]$.
It is not a mapping.

**Fig.S3.15**

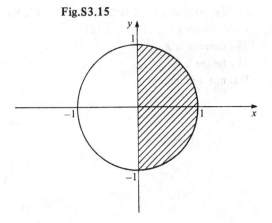

(*e*) $y = 2x - 1$ (Fig. S3.16)
The domain is IR.
The image is IR.
It is a mapping.

**Fig.S3.16**

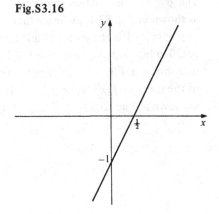

3.9   $f = \{(x, y) \in \mathbb{Q} \times \mathbb{Z} \mid y$ is the least integer with $y \geqslant x\}$ (Fig. S3.17)
The domain is $\mathbb{Q}$.
The image is $\mathbb{Z}$.
It is a mapping.

**Fig.S3.17**

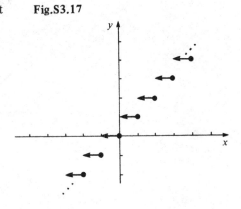

$f = \{(x, y) \in \mathbb{Z} \times \mathbb{Q} \mid x \text{ is the least}$
integer with $x \geqslant y\}$ (Fig. S3.18)
The domain is $\mathbb{Z}$.
The image is $\mathbb{Q}$.
It is not a mapping.

**Fig.S3.18**

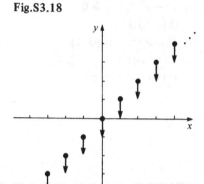

3.10    The graph of the relation $y = |x|$ is
as shown in Fig. S3.19, whence that
of $y = |x| - 1$ is as shown in Fig.
S3.20, whence that of $y = |x - 1| - 1$
is as shown in Fig. S3.21, whence that
of the function $f(x) = ||x - 1| - 1|$ is
as shown in Fig. S3.22.

**Fig.S3.19**

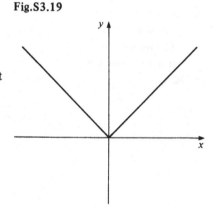

**Fig.S3.20**

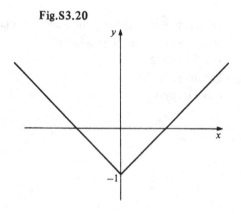

**Fig.S3.21**

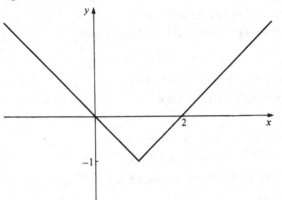

**Fig.S3.22**

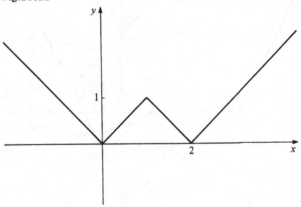

*3.11*   We have
$$2x^2 + 6x + 7 \leqslant x + 5 \Leftrightarrow 2x^2 + 5x + 2 \leqslant 0$$
$$\Leftrightarrow (2x + 1)(x + 2) \leqslant 0,$$
and so the required set is $[-2, -\frac{1}{2}]$.

*3.12*   For every $x \in \mathbb{R}$ define
$$g(x) = \tfrac{1}{2}[f(x) + f(-x)],$$
$$h(x) = \tfrac{1}{2}[f(x) - f(-x)].$$
Then clearly $(g + h)(x) = g(x) + h(x) = f(x)$   for   every   $x \in \mathbb{R}$   and   so

$g + h = f.$ Moreover,

$$g(-x) = \tfrac{1}{2}[f(-x) + f(x)] = g(x),$$
$$h(-x) = \tfrac{1}{2}[f(-x) - f(x)] = -h(x).$$

**3.13**  We have

$$f[g(x)] = f(2x - 1) = (2x)^2 = 4x^2,$$
$$g[f(x)] = g[(x + 1)^2] = 2(x + 1)^2 - 1 = 2x^2 + 4x + 1,$$

and so

$$f[g(x)] = g[f(x)] \Leftrightarrow 2x^2 - 4x - 1 = 0$$
$$\Leftrightarrow x = 1 \pm \tfrac{1}{2}\sqrt{6}.$$

The set $A \cap B \cap C \cap D$ is the subset shown in Fig. S3.23.

**Fig.S3.23**

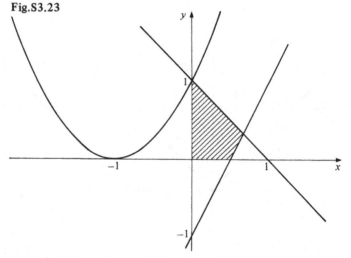

**3.14**  $S$ is the square shown in Fig. S3.24.
$d(x, y)$ is the distance from $(0, 0)$ to $(x, y)$. The maximum distance from $(0, 0)$ to a point $(x, y)$ on $S$ is clearly 1, and the minimum distance is $1/\sqrt{2}$. Since all values between these extremes are attained we have that $\operatorname{Im} d = [1/\sqrt{2}, 1]$.

**Fig.S3.24**

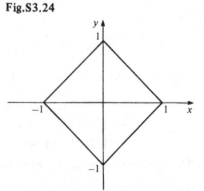

66

## Solutions to Chapter 3

**3.15** $S$ is the shaded region shown in Fig. S3.25.

$f(x, y)$ is the distance from $(0, 0)$ to $(x, y)$. The maximum distance from $(0, 0)$ to a point $(x, y)$ on $S$ is clearly 1, and the minimum distance is $1/\sqrt{2}$. Since all values between these extremes are attained we have that Im $d = [1/\sqrt{2}, 1]$.

**Fig.S3.25**

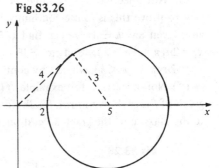

**3.16** $S$ is the circle shown in Fig. S3.26.

$f(x, y)$ is the gradient of the line joining $(0, 0)$ to $(x, y)$. The maximum value occurs when $\tan \vartheta = \frac{3}{4}$ and the minimum value when $\tan \vartheta = -\frac{3}{4}$. Hence we see that Im $f = [-\frac{3}{4}, \frac{3}{4}]$.

**Fig.S3.26**

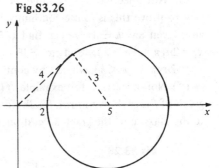

**3.17** $S$ is the circle shown in Fig. S3.27.

$f(x, y)$ is the distance from $(0, 0)$ to $(x, y)$. Now the radius of the circle is 2, and the centre of the circle is at the point $(3, 4)$ which is at distance 5 from the origin. Thus the maximum distance from the origin to a point on $S$ is 7, and the minimum distance is 3. Consequently, Im $f = [3, 7]$.

**Fig.S3.27**

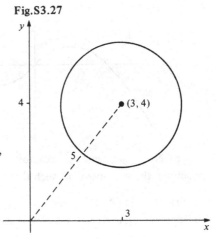

67

3.18
    (*a*) Injective, since $2x_1 + 1 = 2x_2 + 1$ implies $x_1 = x_2$. Not surjective: for example, $\frac{1}{2} \in B = \mathbb{Q}$ but there is no $x \in \mathbb{Z}$ with $f(x) = \frac{1}{2}$.

    (*b*) Injective since $(x_1 - 1, 1) = (x_2 - 1, 1)$ implies $x_1 = x_2$. Not surjective: for example, no $x \in \mathbb{Z}$ maps to $(0, 0)$.

    (*c*) Injective, in fact

$$x_1^3 - x_2^3 = (x_1 - x_2)(x_1^2 + x_1 x_2 + x_2^2)$$

and so if $x_1^3 = x_2^3$ we must have either $x_1 = x_2$ or $x_1^2 + x_1 x_2 + x_2^2 = 0$, and the only solution to the latter equation in $\mathbb{R}$ is $x_1 = x_2 = 0$. Also surjective: given any $y \in B = \mathbb{R}$ we have $f(y^{1/3}) = y$.

    (*d*) Not injective: for example, $f(1) = f(i) = 1$. Not surjective: there is no $c \in \mathbb{C}$ such that $f(c) = -1$ since $|c| \geqslant 0$ for all $c \in \mathbb{C}$.

    (*e*) Not injective: for example, $f(0) = f(\pi)$. Surjective; probably the best way to prove this is to use continuity : the function $x \to x \sin x$ is continuous and, given any $k \in \mathbb{R}$ we can find $x_1 \in \mathbb{R}$ with $f(x_1) > k$ (take for example $x_1 = 2n_1 \pi + \frac{1}{2}\pi > k$), and $x_2 \in \mathbb{R}$ with $f(x_2) < k$ (take for example $x_2 = 2n_2 \pi + \frac{1}{2}\pi < k$), whence by continuity $f(x)$ takes the value $k$.

    (*f*) Not injective: for example, $f(1) = f(0) = 0$. Surjective; in fact the graph of $f$ is as shown in Fig. S3.28, from which we see that every line parallel to the $x$-axis cuts the graph at least once.

    **Fig.S3.28**

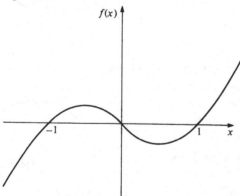

    (*g*) Not injective: for example, $f(0) = f(-1) = 1$. Not surjective: for example, there is no $x \in \mathbb{R}$ such that $x^2 + x + 1 = 0$.

3.19
    (*a*) If $y \in f(A \cap B)$ then $y = f(x)$ for some $x \in A \cap B$. From $x \in A$ we have $y \in f(A)$; and from $x \in B$ we have $y \in f(B)$. Hence $y \in f(A) \cap f(B)$.

(*b*) Since $A \subseteq A \cup B$ it is immediate that $f(A) \subseteq f(A \cup B)$. Similarly, $f(B) \subseteq f(A \cup B)$ and hence $f(A) \cup f(B) \subseteq f(A \cup B)$. To obtain the reverse inclusion, let $y \in f(A \cup B)$. Then $y = f(x)$ for $x \in A \cup B$. If $x \in A$ then $y \in f(A)$, and if $x \in B$ then $y \in f(B)$. In either case, $y \in f(A) \cup f(B)$.

(*c*) Suppose that $f$ is injective. In view of (*a*), we need only prove that $f(A) \cap f(B) \subseteq f(A \cap B)$. Suppose that $y \in f(A) \cap f(B)$. Then $y = f(x_1)$ for $x_1 \in A$ and $y = f(x_2)$ for $x_2 \in B$. Since $f$ is injective by hypothesis, we have $x_1 = x_2 \in A \cap B$ and so $y \in f(A \cap B)$.

Conversely, suppose that $f(A) \cap f(B) \subseteq f(A \cap B)$ for all subsets $A, B$ of $X$. If $f(x_1) = f(x_2) = y$ then from $y = f(x_1) \in f(\{x_1\})$ and $y = f(x_2) \in f(\{x_2\})$ we deduce that $y \in f(\{x_1\} \cap \{x_2\})$, whence it follows that $\{x_1\} \cap \{x_2\} \neq \emptyset$ and hence $x_1 = x_2$.

(*d*) Suppose that $f$ is surjective. If $y \in Y \setminus f(A)$ then there exists $x$ with $f(x) = y$, and clearly $x \in X \setminus A$. Thus $y = f(x) \in f(X \setminus A)$ and consequently $Y \setminus f(A) \subseteq f(X \setminus A)$.

Conversely, suppose that $Y \setminus f(A) \subseteq f(X \setminus A)$ for all subsets $A$ of $X$. Then in particular we have $Y \setminus f(\emptyset) \subseteq f(X \setminus \emptyset)$, i.e. $Y \subseteq f(X)$. It is immediate from this that every $y \in Y$ is of the form $f(t)$ for some $t \in X$, so that $f$ is surjective.

(*e*) Suppose that $f$ is a bijection. Then by (*d*) we have $f(X \setminus A) \supseteq Y \setminus f(A)$ for all subsets $A$. It therefore suffices to show the reverse inclusion. Suppose then that $y \in f(X \setminus A)$. We have $y = f(x)$ for some $x \in X \setminus A$. Now we cannot have $y = f(a)$ for any $a \in A$; for $f$ is injective and $f(x) = f(a)$ would imply $x = a$ where $x \in X \setminus A$ and $a \in A$, a contradiction. Hence we see that $y \notin f(A)$ and consequently $y \in Y \setminus f(A)$.

Conversely, suppose that $f(X \setminus A) = Y \setminus f(A)$ for all subsets $A$ of $X$. Then $f$ is surjective by (*d*). To show that $f$ is also injective, suppose that $x_1 \neq x_2$. Then

$$f(x_1) \in f(X \setminus \{x_2\}) = Y \setminus f(\{x_2\})$$

and so $f(x_1) \notin f(\{x_2\})$ which shows that $f(x_1) \neq f(x_2)$.

3.20     (*a*) Let $f \in (A \times B)^C$ and define $f_1 : C \to A, f_2 : B \to A$ by

         $f_1(x)$ = first component of $f(x)$;

         $f_2(x)$ = second component of $f(x)$.

Define $\varphi : (A \times B)^C \to A^C \times B^C$ by $\varphi(f) = (f_1, f_2)$. Then $\varphi$ is clearly a bijection.

    (*b*) If $f \in (A^B)^C$ then $f : C \to A^B$ and, for each $c \in C, f(c) : B \to A$. Thus

$$(\forall c \in C)(\forall b \in B)\,[f(c)](b) \in A.$$

We can therefore define a mapping $\varphi : (A^B)^C \to A^{B \times C}$ by $f \to \varphi(f)$ where
$$(\forall c \in C)(\forall b \in B) [\varphi(f)](b, c) = [f(c)](b).$$
Clearly, $\varphi$ is a bijection.

(c) Let $f \in A^{B \cup C}$, so that $f : B \cup C \to A$. Let $f_1 : B \to A$ be given by $(\forall b \in B)f_1(b) = f(b)$ and let $f_2 : C \to A$ be given by $(\forall c \in C)f_2(c) = f(c)$. Now define $\varphi : A^{B \cup C} \to A^B \times A^C$ by $\varphi(f) = (f_1, f_2)$. Then $\varphi$ is clearly a bijection when $B \cap C = \emptyset$.

Although $B \cap C = \emptyset$ is a sufficient condition for the existence of a bijection from $A^{B \cup C}$ to $A^B \times A^C$, it is not a necessary condition. For example, take $A = B = C = \{1\}$; we have $A^{B \cup C} = \{f\}$ where $f$ is described by $1 \to 1$, $A^B = \{f\} = A^C$, and there is a bijection from $\{f\}$ to $\{f\} \times \{f\}$.

3.21    (a) $x \to x + 1$.

(b) $x \to 2x$.

(c) Let $A$ be the point $(1, -1)$ and let $B$ be the point $(-1, 1)$. For $x_1 \in [0, 1[$ let $f(x_1)$ be the point of intersection on the $y$-axis of the line through $A$ and $x_1$; and for $x_2 \in ]-1, 0]$ let $f(x_2)$ be the point of intersection on the $y$-axis of the line through $B$ and $x_2$. This describes a bijection $f : ]-1, 1[ \to \mathbb{R}$; as is readily verified, for $x_1 \in [0, 1[$ we have $f(x_1) = x_1/(1 - x_1)$ and for $x_2 \in ]-1, 0]$ we have $f(x_2) = x_2/(1 + x_2)$. See Fig. S3.29.

**Fig. S3.29**

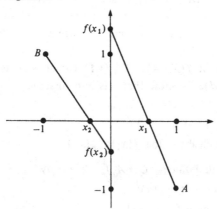

(d) Define $f : [0, 1] \to [0, 1[$ by
$$f(x) = \begin{cases} \frac{1}{2}x & \text{if} \quad x = 1/2^n; \\ x & \text{otherwise.} \end{cases}$$

See Fig. S3.30.

**Fig.S3.30**

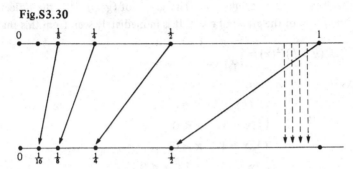

(*e*) Note that in (*d*) the formula for $f$ can be extended to negative values of $x$ thereby providing a bijection from $[-1, 1]$ to $]-1, 1[$. Composing this with the bijection in (*c*), we obtain a bijection from $[-1, 1]$ to $\mathbb{R}$.

*3.22*  Using the formulae given, we have

$$g[f(x)] = \begin{cases} g(4x+1) & x \geqslant 0; \\ g(x) & x < 0; \end{cases}$$

$$= \begin{cases} 12x+3 & x \geqslant 0; \\ x+3 & x < 0. \end{cases}$$

The graph of $g \circ f$ is as shown in Fig. S3.31.

**Fig.S3.31**

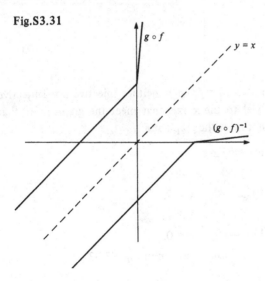

Since each line parallel to the $x$-axis meets the graph exactly once it follows that $g \circ f$ is a bijection. The graph of $(g \circ f)^{-1}$ is the reflection in the line $y = x$ of the graph of $g \circ f$. It is immediately seen from this that

$$(g \circ f)^{-1}(x) = \begin{cases} x - 3 & x < 3; \\ \frac{1}{12}(x - 3) & x \geqslant 3. \end{cases}$$

As for $f \circ g$, we have

$$f[g(x)] = \begin{cases} f(3x) & x \geqslant 0; \\ f(x + 3) & x < 0, \end{cases}$$

$$= \begin{cases} 12x + 1 & x \geqslant 0; \\ 4x + 13 & -3 \leqslant x < 0; \\ x + 3 & x < -3. \end{cases}$$

The graph of $f \circ g$ is as shown in Fig. S3.32.

**Fig.S3.32**

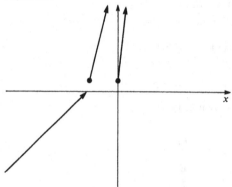

It is clear from this that $f \circ g$ is neither injective nor surjective (since there are lines parallel to the $x$-axis that meet the graph more than once, and others that do not meet the graph at all).

3.23     Applying the formulae we have

$$f[g(x)] = \begin{cases} f(x) & x \geqslant 0; \\ f(x - 1) & x < 0, \end{cases}$$

$$= \begin{cases} 1 - x & x \geqslant 0; \\ (x - 1)^2 & x < 0. \end{cases}$$

The graph of $f \circ g$ is therefore as shown in Fig. S3.33.

**Fig.S3.33**

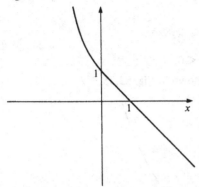

Since each line parallel to the $x$-axis meets this graph precisely once, it follows that $f \circ g$ is a bijection. The formula for $(f \circ g)^{-1}$ is

$$(f \circ g)^{-1}(x) = \begin{cases} 1-x & x \leqslant 1; \\ 1-\sqrt{x} & x > 1. \end{cases}$$

As for $g \circ f$ we have

$$g[f(x)] = \begin{cases} g(1-x) & x \geqslant 0; \\ g(x^2) & x < 0, \end{cases}$$

$$= \begin{cases} -x & x > 1; \\ 1-x & 0 \leqslant x \leqslant 1; \\ x^2 & x < 0. \end{cases}$$

The graph of $g \circ f$ is as shown in Fig. S3.34.

**Fig.S3.34**

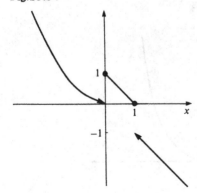

It is clear from this that $g \circ f$ is neither injective nor surjective.

3.24    Since $f(x) = x|x|$ we have

$$f(x) = \begin{cases} x^2 & \text{if} \quad x \geqslant 0; \\ -x^2 & \text{if} \quad x < 0. \end{cases}$$

The graph of $f$ is therefore as shown in Fig. S3.35.

**Fig.S3.35**

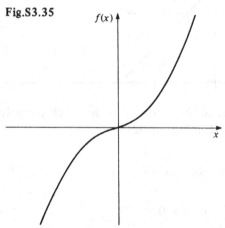

Since every line parallel to the $x$-axis meets the graph exactly once, $f$ is a bijection.

For $x \geqslant 0$ we have $y = x^2$ and so $x = \sqrt{y}$; and for $x < 0$ we have $y = -x^2$ and so $x = -\sqrt{(-y)}$. Hence $f^{-1}$ is given by

$$f^{-1}(x) = \begin{cases} \sqrt{x} & \text{if} \quad x \geqslant 0; \\ -\sqrt{(-x)} & \text{if} \quad x < 0. \end{cases}$$

The graph of $g$ is as shown in Fig. S3.36.

**Fig.S3.36**

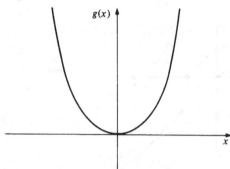

$g$ is not surjective : for example, the line $y = -1$ does not meet the graph.

$g$ is not injective : for example, the line $y = 1$ meets the graph twice.

Hence $g$ is not a bijection.

3.25    In general, the cubic $x \to x^3 + ax^2 + bx + c$ has a graph of the form shown in Fig. S3.37.

**Fig.S3.37**

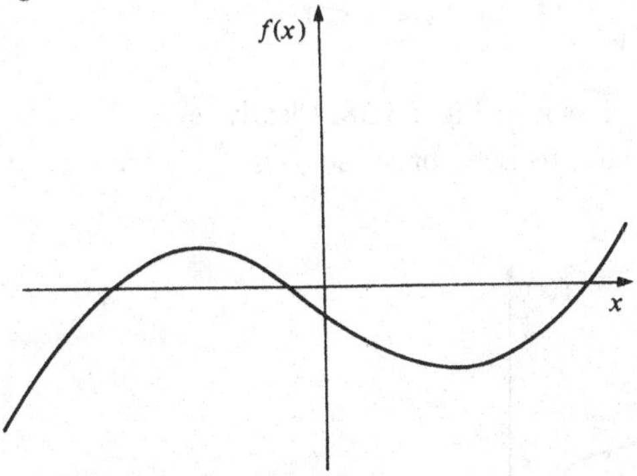

For this function to be a bijection, it is necessary and sufficient that every line parallel to the $x$-axis meets the graph precisely once. This is so if and only if the gradient of the function is greater than or equal to 0. Now the derivative is $3x^2 + 2ax + b$, which is a quadratic. The condition that $f$ be a bijection is therefore $(\forall x \in \mathbb{R}) 3x^2 + 2ax + b \geqslant 0$. Now the quadratic $x \to 3x^2 + 2ax + b$ attains its minimum value when $6x + 2a = 0$, i.e. when $x = -\frac{1}{3}a$, this minimum value then being $-\frac{1}{3}a^2 + b$. This is greater than or equal to 0 if and only if $a^2 \leqslant 3b$.

3.26    If $f(x_1) = f(x_2)$ then from

$$\frac{1}{1+x_1^2} = \frac{1}{1+x_2^2}$$

we deduce $x_1^2 = x_2^2$ whence $x_1 = \pm x_2$. Since $x_1, x_2 \in \mathbb{R}_+$ we have $x_1 = x_2$ and so $f$ is injective.

Since $f$ is injective, the assignment $x \to f(x)$ defines a bijection $\vartheta : \mathbb{R}_+ \to$ Im $f$ and the inverse of this bijection is given by

$$\vartheta^{-1}(x) = \sqrt{\left(\frac{1}{x} - 1\right)}.$$

To find two distinct mappings $g, h : \mathbb{R}_+ \to \mathbb{R}_+$ such that $g \circ f = h \circ f = \mathrm{id}_{\mathbb{R}_+}$, it suffices to find two mappings which agree with $\vartheta^{-1}$ on Im $f$. Thus, for example, we can consider

$$g(x) = \begin{cases} 0 & \text{if } x = 0 \text{ or } x > 1; \\ \sqrt{\left(\dfrac{1}{x} - 1\right)} & \text{if } 0 < x \leqslant 1. \end{cases}$$

$$h(x) = \begin{cases} 1 & \text{if } x = 0 \text{ or } x > 1; \\ \sqrt{\left(\dfrac{1}{x} - 1\right)} & \text{if } 0 < x \leqslant 1. \end{cases}$$

3.27    The graph of $f$ is as shown in Fig. S3.38. Clearly, every line parallel to the $x$-axis meets the graph precisely once, so $f$ is a bijection. The derivative

**Fig.S3.38**

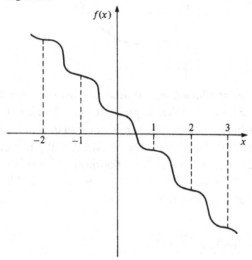

$f'$ of $f$ is given by

$$f'(x) = -\pi \sin \left[\pi(x - [\![x]\!])\right].$$

Since, for example, $f'(0) = f'(1) = 0$ we see that $f'$ is not a bijection.

3.28    The mapping can be described by

$$\begin{array}{ccccccccc} 1 & 2 & 3 & 4 & 5 & 6 & 7 & \ldots \\ \downarrow & \downarrow & \downarrow & \downarrow & \downarrow & \downarrow & \downarrow & \\ 0 & 1 & -1 & 2 & -2 & 3 & -3 & \ldots \end{array}$$

which suggests that $f$ is a bijection whose inverse is given by

$$f^{-1}(m) = \begin{cases} 2m & \text{if } m > 0; \\ 2|m| + 1 & \text{if } m \leqslant 0. \end{cases}$$

Now with this definition of $f^{-1}$ we have

$$f[f^{-1}(m)] = \begin{cases} (-1)^{2m} [\![m]\!] & \text{if } m > 0; \\ (-1)^{2|m|+1}|m| & \text{if } m \leqslant 0, \end{cases}$$

$$= \begin{cases} m & \text{if } m > 0; \\ -|m| & \text{if } m \leqslant 0, \end{cases}$$

$$= m.$$

Consequently we see that $f \circ f^{-1} = \mathrm{id}_{\mathbb{Z}}$ and so $f$ is certainly surjective. Since $f$ is clearly injective by definition, it follows that $f$ is indeed a bijection with $f^{-1}$ as described above.

3.29  Since $g(x) = 3 + 4x$ the formula $g^n(x) = (4^n - 1) + 4^n x$ certainly holds for $n = 1$. For the inductive step, suppose that $g^n(x) = (4^n - 1) + 4^n x$. Then

$$g^{n+1}(x) = g[g^n(x)] = 3 + 4[(4^n - 1) + 4^n x]$$
$$= 3 + 4^{n+1} - 4 + 4^{n+1}x$$
$$= (4^{n+1} - 1) + 4^{n+1}x.$$

For the last part, we have

$$g^{-n}(x) = [g^n(x)]^{-1} = \frac{1}{4^n}[x - (4^n - 1)]$$
$$= 4^{-n}x + (4^{-n} - 1),$$

whence we see that the formula holds also for negative integers $n$.

3.30  For $n = 1$ the formula is trivial, and for $n = 2$ it reads

$$|A_1 \cup A_2| = |A_1| + |A_2| - |A_1 \cap A_2|.$$

For the inductive step, suppose that the result holds for $n$ subsets. Then using the result for two subsets we have

$$\left| \bigcup_{i=1}^{n+1} A_i \right| = \left| \left( \bigcup_{i=1}^{n} A_i \right) \cup A_{n+1} \right|$$

$$= \left| \bigcup_{i=1}^{n} A_i \right| + |A_{n+1}| - \left| \left( \bigcup_{i=1}^{n} A_i \right) \cap A_{n+1} \right|$$

$$= \sum_{i=1}^{n+1} |A_i| - \sum_{i<j} |A_i \cap A_j| + \cdots + (-1)^n \left| \bigcap_{i=1}^{n} A_i \right|$$

$$- |(A_1 \cap A_{n+1}) \cup \cdots \cup (A_n \cap A_{n+1})|.$$

Using the formula again, this last component (which is a union of $n$ sets) can be written

$$-\left(\sum_{i=1}^{n}|A_i\cap A_{n+1}|-\sum_{p<q}|A_p\cap A_q\cap A_{n+1}|+\cdots\right)$$

which, when taken with the other terms, gives the required formula for $n+1$. This completes the inductive step.

If $|A|=m$ and $|B|=n$ then in defining a mapping from $A$ to $B$ we have, for every element $x$ of $A$, $n$ choices of image in $B$. Thus there are $n^m$ mappings from $A$ to $B$.

Suppose now that $A=\{a_1,a_2,\ldots,a_m\}$ and that $f:A\to B$ is an injection. There are $n$ possible images of $a_1$, then $n-1$ possible images of $a_2$, and so on. Hence there are

$$n(n-1)(n-2)\cdots(n-m+1)=\frac{n!}{m!}$$

injections from $A$ to $B$.

To determine the number of surjections, let $A=\{a_1,a_2,\ldots,a_m\}$ and $B=\{b_1,b_2,\ldots,b_n\}$. For each $b_i\in B$ let

$$A_i=\{f:A\to B\mid b_i\notin f(A)\}.$$

Then the number of surjections from $A$ to $B$ is the total number of mappings from $A$ to $B$ less $|\cup_{i=1}^{n}A_i|$, i.e. it is $n^m-|\cup_{i=1}^{n}A_i|$. It is at this stage that we require the general formula established above. To apply this, we observe that $|A_i|$ is the number of mappings from $A$ to $B\backslash\{b_i\}$, $|A_i\cap A_j|$ is the number of mappings from $A$ to $B\backslash\{b_i,b_j\}$, etc. Thus we have that

$$|A_i|=(n-1)^m,\quad |A_i\cap A_j|=(n-2)^m,\ldots$$

Since we can remove $i$ elements of $B$ in precisely $\binom{n}{i}$ ways, each sum $\Sigma\mid\cap_{i\,\text{terms}}A_k\mid$ contains $\binom{n}{i}$ terms equal to $(n-i)^m$ and the result follows.

3.31    Let $A=B=C=[0,1]$ and let each mapping to be given by $f(x)=\frac{1}{2}x$. Then $f$ is an injection which is not a bijection and which has the required properties.

For another example, consider $A=[0,1]\times[0,1]$, $B=[0,1]\times[0,\frac{2}{3}]$, $C=[0,\frac{3}{4}]\times[0,1]$ and $f:A\to B, g:B\to C, h:C\to A$ given by

$$f(x,y)=(x,\tfrac{1}{2}y),\quad g(x,y)=(y,x),\quad h(x,y)=(x,y).$$

3.32    Use calculus to determine the graph of $f$ (Fig. S3.39). From this graph we see that Im $f=[-2,2]$ and that the required interval $A$ is $[-1,1]$. To find $g^{-1}$ we set $y=4x/(x^2+1)$, so that $yx^2-4x+y=0$, and solve this quadratic

**Fig.S3.39**

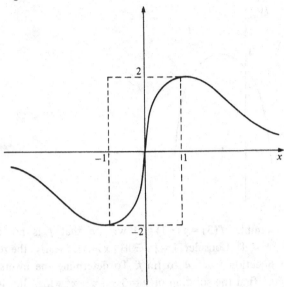

for $x$ in terms of $y$. We get

$$x = \begin{cases} \dfrac{4 \pm \sqrt{(16 - 4y^2)}}{2y} = \dfrac{2 \pm \sqrt{(4 - y^2)}}{y} & \text{if} \quad y \neq 0; \\ 0 & \text{if} \quad y = 0. \end{cases}$$

This suggests either that

$$g^{-1}(y) = \begin{cases} \dfrac{2 + \sqrt{(4 - y^2)}}{y} & \text{if} \quad y \neq 0; \\ 0 & \text{if} \quad y = 0, \end{cases}$$

or that

$$g^{-1}(y) = \begin{cases} \dfrac{2 - \sqrt{(4 - y^2)}}{y} & \text{if} \quad y \neq 0; \\ 0 & \text{if} \quad y = 0. \end{cases}$$

The first possibility is excluded since, for $y \in [-1, 1]$, we have

$$\frac{2 + \sqrt{(4 - y^2)}}{y} \notin [-2, 2].$$

3.33    The graph of $f$ is shown in Fig. S3.40.

Fig.S3.40

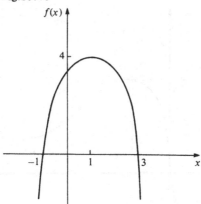

Since, for example, $f(3) = f(-1) = 0$ we see that $f$ is not injective. Im $f = \{x \in \mathbb{R} \mid x \leqslant 4\}$. Consider $A = \{x \in \mathbb{R} \mid x \geqslant 1\}$. Clearly, the restriction of $f$ to $A$ is a bijection from $A$ to Im $f$. To determine the inverse of this bijection we must find the solution of $y = 3 + 2x - x^2$ which lies in $A$. The solutions of this quadratic equation are $x = 1 \pm \sqrt{(4-y)}$ of which only $1 + \sqrt{(4-y)}$ lies in $A$. Hence the required inverse is described by $x \to 1 + \sqrt{(4-x)}$.

3.34   Note from the definition of $f$ that $f(n)$ is divisible by $p$ if and only if $n$ is divisible by $p$. Thus if $f(n_1) = f(n_2)$ then either $f(n_1)$ is divisible by $p$, in which case $n_1 = f(n_1) - p = f(n_2) - p = n_2$, or $f(n_1)$ is not divisible by $p$, in which case $n_1 = f(n_1) = f(n_2) = n_2$. Hence $f$ is an injection.

   To show that $f$ is also surjective, suppose that $k \in \mathbb{Z}$. If $k$ is not divisible by $p$ then $f(k) = k$; and if $k$ is divisible by $p$ then so is $k - p$ whence $f(k - p) = (k - p) + p = k$. Hence $f$ is a surjection.

   As for $f^{-1} \colon \mathbb{Z} \to \mathbb{Z}$, this is given by

$$f^{-1}(n) = \begin{cases} n & \text{if } n \text{ is not divisible by } p; \\ n - p & \text{if } n \text{ is divisible by } p. \end{cases}$$

3.35   It suffices to find $g \colon \mathbb{R} \to \mathbb{R}$ such that $f \circ g = g \circ f = \mathrm{id}_{\mathbb{R}}$. Then $f$ is a bijection with $f^{-1} = g$. Now if $y = x^4$ we have $x = y^{1/4}$ so we can define

$$g(x) = x^{1/4} \quad \text{if} \quad x \geqslant 0.$$

Also, if $y = x(2 - x)$ then $x^2 - 2x + y = 0$ and we have $x = 1 \pm \sqrt{(1-y)}$.

80

Since $1 + \sqrt{(1-y)} > 0$ we must choose $x = 1 - \sqrt{(1-y)}$ and define

$$g(x) = 1 - \sqrt{(1-x)} \quad \text{if} \quad x < 0.$$

Then it is easy to check that $f[g(x)] = x$ and $g[f(x)] = x$ for every $x \in \mathbb{R}$, so that $g$ is the inverse of $f$.

3.36  (a) Since we have $1 = f(2) < f(3) = 2$ the result holds for $i = 2$. By way of applying the second principle of induction, suppose that the result holds for $2 \leqslant i \leqslant n$. Then we have

$$f(n+1) = f(n) + f(n-1)$$
$$< f(n+1) + f(n)$$
$$= f(n+2),$$

whence we see that it also holds for $n + 1$.

(b) If $f[f(i)] = f(i)$ then, by (a), if $i > 2$ we must have $f(i) = i$ and the only possibility is $i = 5$; and if $i \leqslant 2$ then it is readily seen that the possibilities are $i = 0, 1, 2$.

(c) For $n = 1$ we have $f_5 = 5$. Assume by way of induction that the result holds for $n$. Then

$$f_{5(n+1)} = f_{5n+5} = f_{5n+4} + f_{5n+3}$$
$$= 2f_{5n+3} + f_{5n+2}$$
$$= 3f_{5n+2} + 2f_{5n+1}$$
$$= 5f_{5n+1} + 3f_{5n},$$

whence it also holds for $n + 1$.

(d) Since $f_3 = 2 = 1 + f_1$, the result holds for $n = 1$. Assume by way of induction that it holds for $n$, so that $f_{n+2} = 1 + \sum_{i=1}^{n} f_i$. Then we have

$$f_{n+3} = f_{n+2} + f_{n+1}$$

$$= 1 + \sum_{i=1}^{n} f_i + f_{n+1}$$

$$= 1 + \sum_{i=1}^{n+1} f_i,$$

whence it also holds for $n + 1$.

(e) The result is clearly true for $n = 1$. Assume by way of induction that the result holds for $n$, so that $f_{n-1}f_{n+1} - f_n^2 = (-1)^n$. Then we have

$$f_n f_{n+2} - f_{n+1}^2 = f_n(f_n + f_{n+1}) - f_{n+1}^2$$
$$= f_n^2 + f_n f_{n+1} - f_{n+1}(f_n + f_{n-1})$$

$$= f_n^2 - f_{n+1}f_{n-1}$$
$$= (-1)^n(-1)$$
$$= (-1)^{n+1},$$

whence it also holds for $n + 1$.

($f$) The result clearly holds for $n = 1$. By way of applying the second principle of induction, suppose that it holds for all $n \leqslant k$. Then we have

$$2^{k+1}\sqrt{5}\, f_{k+1} = 2^{k+1}\sqrt{5}\, f_k + 2^{k+1}\sqrt{5}\, f_{k-1}$$
$$= 2(1 + \sqrt{5})^k - 2(1 - \sqrt{5})^k$$
$$+ 4(1 + \sqrt{5})^{k-1} - 4(1 - \sqrt{5})^{k-1}$$
$$= (1 + \sqrt{5})^{k-1}(2 + 2\sqrt{5} + 4)$$
$$- (1 - \sqrt{5})^{k-1}(2 - 2\sqrt{5} + 4)$$
$$= (1 + \sqrt{5})^{k+1} - (1 - \sqrt{5})^{k+1},$$

whence we see that the result holds also for $n = k + 1$.

3.37    Suppose that $g(1) = 3$ and that $f \circ g = g \circ f$. Then we have $f[g(1)] = f(3) = 4$ and hence $g[f(1)] = 4$, i.e. $g(2) = 4$. Now $f[g(2)] = f(4) = 1$ so $g[f(2)] = 1$, i.e. $g(3) = 1$. Finally, $f[g(3)] = f(1) = 2$ so $g[f(3)] = 2$, i.e. $g(4) = 2$. Thus we see that $g$ is determined uniquely and is given by

$$g(x) = x + 2 \quad (\text{mod } 4).$$

A similar argument shows that if $h(1) = 1$ then $h$ is uniquely determined, and $h = \text{id}_X$.

3.38    ($a$) Clearly such a mapping $g$ must satisfy $\alpha[g(0)] = 0$. But there is no $m \in \text{IN}$ for which $\alpha(m) = 0$. Hence such a mapping $g$ cannot exist. For every $p \in \text{IN}$ let $k_p : \text{IN} \to \text{IN}$ be given by

$$k_p(n) = \begin{cases} n - 1 & \text{if} \quad n \geqslant 1; \\ p & \text{if} \quad n = 0. \end{cases}$$

Then for every $m \in \text{IN}$ we have

$$k_p[\alpha(m)] = k_p(m + 1) = m,$$

so that each mapping $k_p$ is such that $k_p \circ \alpha = \text{id}_{\text{IN}}$. Clearly there are infinitely many such $k_p$.

($b$) Suppose that there exists $f : \text{IN} \to \text{IN}$ such that $f \circ \beta = \text{id}_{\text{IN}}$. Then we have

$$3 = f[\beta(3)] = f(1)$$
$$2 = f[\beta(2)] = f(1),$$

and from this contradiction we conclude that no such $f$ can exist. For every $p \in \mathbb{N}$ let $k_p : \mathbb{N} \to \mathbb{N}$ be given by

$$k_p(n) = \begin{cases} 2n + 1 & \text{if} \quad n \neq p; \\ 2p & \text{if} \quad n = p. \end{cases}$$

For $m \neq p$ we have

$$\beta\,[k_p(m)] = \beta(2m + 1) = m,$$

and for $m = p$ we have

$$\beta\,[k_p(m)] = \beta(2p) = p = m.$$

Thus $\beta \circ k_p = \mathrm{id}_{\mathbb{N}}$. Clearly there are infinitely many such $k_p$.

3.39    Suppose that there exists $x \in A$ such that $f(x) = X$ where, by definition, $X = \{a \in A \mid a \notin f(a)\}$. Then if $x \in f(x)$ we have $x \in X$ whence the contradiction $x \notin f(x)$. Also, if $x \notin f(x)$ then we have $x \notin X$ whence the contradiction $x \in f(x)$. We conclude that there can be no $x \in A$ with $f(x) = X$. It is immediate from this observation that $f$ cannot be surjective. There does exist, however, an injection $f : A \to \mathbf{P}(A)$: for example, $x \to f(x) = \{x\}$.

3.40    (a) It is helpful to compute a few of the $I_k$. For example,

$$I_1 = \{1\}, \quad I_2 = \{2, 3\}, \quad I_3 = \{4, 5, 6\}, \quad I_4 = \{7, 8, 9, 10\}.$$

Now by the definition of $I_k$ we have $\frac{1}{2}k(k + 1) \in I_k$ and $\frac{1}{2}k(k + 1) - k = \frac{1}{2}k(k - 1) \notin I_k$. It is clear from this that $I_k$ consists of the $k$ elements

$$\tfrac{1}{2}k(k + 1) - k + 1, \ldots, \tfrac{1}{2}k(k + 1).$$

Consequently we see that the $I_k$ form a partition of $\mathbb{N}^*$.

(b) We observe that

$$\tfrac{1}{2}(m + n - 2)(m + n - 1) + m = \tfrac{1}{2}(m + n)(m + n - 1) + 1 - n$$

and that this belongs to $I_{m+n-1}$ since the largest integer in $I_{m+n-1}$ is $\frac{1}{2}(m + n)(m + n - 1)$ and $I_{m+n-1}$ contains $m + n - 1$ integers. Thus $f(m, n) \in I_{m+n-1}$.

Since $f(p, q) \in I_{p+q-1}$ and the $I_k$ form a partition, we have

$$f(p, q) \in I_{m+n-1} \Leftrightarrow I_{p+q-1} = I_{m+n-1}$$
$$\Leftrightarrow p + q - 1 = m + n - 1$$
$$\Leftrightarrow p + q = m + n.$$

To show that $f$ is injective, suppose that $f(p, q) = f(m, n)$. Then $p + q = m + n$ and so

$$\tfrac{1}{2}(p+q-2)(p+q-1)+p=f(p,q)$$
$$=f(m,n)$$
$$=\tfrac{1}{2}(m+n-2)(m+n-1)+m$$
$$=\tfrac{1}{2}(p+q-1)(p+q-1)+m.$$

Consequently we have that $m=p$, and hence also $q=n$. Thus $(p,q)=(m,n)$ and so $f$ is injective.

(c) We note that if $1\leqslant r\leqslant k$ then

$$f(r,k+1-r)\in I_{r+(k+1-r)-1}=I_k.$$

But $f$ is injective, $I_k$ contains $k$ elements, and there are $k$ values of $r$; therefore we must have that every element of $I_k$ is the image of some element $(r,k+1-r)$ under $f$. Since this is true for every $I_k$ in the partition, it follows that $f$ is also surjective.

3.41    Consider the mapping $f:S\to S$ defined by

$$f(x)=\begin{cases} 1/x & \text{if}\quad x\in\,]-1,1[\ \text{and}\ x\neq 0; \\ 0 & \text{if}\quad x=0; \\ -1/x & \text{if}\quad x\notin\,]-1,1[. \end{cases}$$

It is readily seen that for every $x\in S$ we have $f[f(x)]=-x$. The graph of $f$ is shown in Fig. S3.41.

**Fig.S3.41**

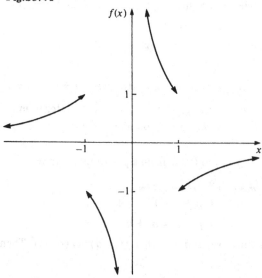

## Solutions to Chapter 3

**3.42**  Since $g \circ f$ is a bijection we have that $g$ is surjective; and since $h \circ g$ is a bijection we have that $g$ is injective. Thus $g$ is a bijection. Since $g \circ f = k$ where $k$ is a bijection, we thus have that $f = g^{-1} \circ k$ is also a bijection. Likewise, since $h \circ g = m$ where $m$ is a bijection, we have that $h = m \circ g^{-1}$ is also a bijection.

**3.43**  Let $\alpha = \text{hcf}(a, b)$. Then we have $a = a'\alpha$, $b = b'\alpha$ and $a/b = a'/b'$, the latter quotient being 'in its lowest terms' in the sense that $\text{hcf}(a', b') = 1$. Similarly, if $\beta = \text{hcf}(c, d)$ then $c = c'\beta$, $d = d'\beta$ and $c/d = c'/d'$ the latter being in its lowest terms. Thus, if $a/b = c/d$ we have $a'/b' = c'/d'$ so that either $a' = c'$, $b' = d'$ or $a' = -c', b' = -d'$. It follows that

$$\left| \frac{a+b}{\alpha} \right| = |a' + b'| = |c' + d'| = \left| \frac{c+d}{\beta} \right|.$$

The first part of the question is precisely the condition that is necessary to ensure that the given prescription defines a mapping from $\mathbb{Q}_+$ to itself. This mapping is not a bijection. For example, it fails to be injective: we have $f(2/3) = 5/1 = f(3/2)$.

**3.44**  (a) Take, for example, $f = \text{id}_{\mathbb{R}}$ and $g = -\text{id}_{\mathbb{R}}$. Then $(f + g)(x) = 0$ for every $x \in \mathbb{R}$ and so $f + g$ is not a bijection.

Consider now $f = \text{id}_{\mathbb{R}}$ and define $g : \mathbb{R} \to \mathbb{R}$ by

$$g(x) = \begin{cases} \dfrac{1}{x} & \text{if } x \neq 0; \\ 0 & \text{if } x = 0. \end{cases}$$

Then $g$ is a bijection; its graph is shown in Fig. S3.42.

**Fig.S3.42**

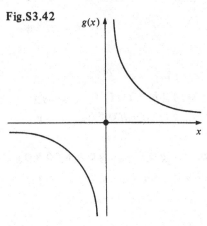

Now

$$(f \cdot g)(x) = f(x)g(x) = \begin{cases} x \cdot \dfrac{1}{x} = 1 & \text{if} \quad x \neq 0; \\ 0 \cdot 0 = 0 & \text{if} \quad x = 0, \end{cases}$$

so $f \cdot g$ is not a bijection. For this $f$ and $g$ we also have

$$(f + g)(x) = \begin{cases} x + \dfrac{1}{x} & \text{if} \quad x \neq 0; \\ 0 & \text{if} \quad x = 0. \end{cases}$$

The equation $x + 1/x = 1$ has no solution in IR, so there is no $x \in$ IR such that $(f + g)(x) = 1$. Thus $f + g$ is not a bijection.

(b) Suppose that $f$ is a bijection. Then for $\lambda \neq 0$ we have

$$(\lambda f)(x_1) = (\lambda f)(x_2) \Rightarrow \lambda f(x_1) = \lambda f(x_2)$$
$$\Rightarrow f(x_1) = f(x_2)$$
$$\Rightarrow x_1 = x_2$$

and so $\lambda f$ is injective. Also, since $f$ is surjective, given any $y \in$ IR, there exists $t \in$ IR with $f(t) = y/\lambda$. Then $(\lambda f)(t) = y$ and hence $\lambda f$ is also surjective. This shows that if $f$ is a bijection then so is $\lambda f$ for every $\lambda \neq 0$. Suppose now that $\lambda f$ is a bijection with $\lambda \neq 0$. Applying the above result to $\lambda f$ and $\mu = 1/\lambda$, it follows that $\mu(\lambda f) = (1/\lambda)(\lambda f) = f$ is also a bijection.

(c) Consider, for example, the mappings $f, g :$ IR $\rightarrow$ IR given by $f(x) = x^2$, $g(x) = x + 1$. We have

$$[fg](x) = f[g(x)] - g[f(x)]$$
$$= (x + 1)^2 - (x^2 + 1)$$
$$= 2x,$$

so $[fg]$ is a bijection.

(d) We have

$$[[fg]h] = [(f \circ g - g \circ f)h]$$
$$= (f \circ g - g \circ f) \circ h - h \circ (f \circ g - g \circ f)$$
$$= f \circ g \circ h - g \circ f \circ h - h \circ f \circ g + h \circ g \circ f$$

and similarly

$$[[gh]f] = g \circ h \circ f - h \circ g \circ f - f \circ g \circ h + f \circ h \circ g,$$
$$[[hf]g] = h \circ f \circ g - f \circ h \circ g - g \circ h \circ f + g \circ f \circ h.$$

Adding these together, we obtain the required result.

86

**3.45**     (*a*) We have that

$$(a + ib)R(c + id) \Leftrightarrow \frac{a}{\sqrt{(a^2 + b^2)}} = \frac{c}{\sqrt{(c^2 + d^2)}}$$

from which it is immediate that $R$ is an equivalence relation on $\mathbb{C}^*$. Now $R$ can also be expressed in the form

$$zRw \Leftrightarrow \frac{\text{Re } z}{|z|} = \frac{\text{Re } w}{|w|}$$

from which we see that $zRw \Rightarrow \bar{z}R\bar{w}$. We now observe that $(z/|z|)Rz$ for every $z \in \mathbb{C}^*$, where $w = z/|z|$ lies on the circle $|w| = 1$ and so can be written $w = e^{i\vartheta}$. Since Re $w/|w| = \cos \vartheta$, it is readily seen that in the Argand diagram the $R$-class of $z \in \mathbb{C}^*$ consists of a pair of half-lines emanating from the origin (with, of course, the origin deleted), one of these lines passing through $z$ and the other through $\bar{z}$ (Fig. S3.43).

**Fig.S3.43**

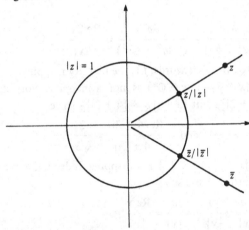

(*b*) Suppose that $x = a + ib$ and $y = c + id$ in $\mathbb{C}^*$ are such that $xRy$. Then we have

$$\frac{a^2}{a^2 + b^2} = \frac{c^2}{c^2 + d^2}$$

from which it follows that

$$\frac{a^2 - b^2}{a^2 + b^2} = \frac{2a^2}{a^2 + b^2} - 1 = \frac{2c^2}{c^2 + d^2} - 1 = \frac{c^2 - d^2}{c^2 + d^2}.$$

Now $x^2 = a^2 - b^2 + 2iab$ and $|x^2| = |x|^2 = a^2 + b^2$ and hence

$$x^2 R y^2 \Leftrightarrow \frac{\mathrm{Re}\, x^2}{|x^2|} = \frac{\mathrm{Re}\, y^2}{|y^2|}$$

$$\Leftrightarrow \frac{a^2 - b^2}{a^2 + b^2} = \frac{c^2 - d^2}{c^2 + d^2}.$$

Consequently we see that $xRy \Rightarrow x^2 R y^2$.

It is clear from this that $\{([x]_R, [x^2]_R) \mid x \in \mathbb{C}^*\}$ describes a mapping $f : U \to U$. As observed above, we can represent every $R$-class as $[e^{i\vartheta}]_R$ where $0 \leqslant \vartheta \leqslant \pi$. Using this representation, we see that this mapping $f$ is surjective: for clearly $[e^{i\vartheta}]_R = f([e^{i\vartheta/2}]_R)$. It is also injective since if $[e^{i\vartheta}]_R = [e^{i\varphi}]_R$ then, by the definition of $R$,

$$\cos 2\vartheta = \mathrm{Re}\, e^{2i\vartheta} = \mathrm{Re}\, e^{2i\varphi} = \cos 2\varphi$$

whence $\vartheta = \varphi$ since each belongs to $[0, \pi]$, and consequently $[e^{i\vartheta}]_R = [e^{i\varphi}]_R$.

(c) (i) $f = \{([x]_R, [2x]_R) \mid x \in \mathbb{C}^*\}$ is also a mapping. Indeed, if $x = a + ib$ then

$$\frac{\mathrm{Re}\, x}{|x|} = \frac{a}{\sqrt{(a^2 + b^2)}} = \frac{2a}{\sqrt{(4a^2 + 4b^2)}} = \frac{\mathrm{Re}\, 2x}{|2x|},$$

so that $[x]_R = [2x]_R$. In other words, $f$ is the identity mapping.

(ii) $f = \{([x]_R, [x + 2]_R) \mid x \in \mathbb{C}^*\}$ is not a mapping. For example, we have $[1 + i]_R = [2 + 2i]_R$ but $[3 + i]_R \neq [4 + 2i]_R$ since

$$\frac{\mathrm{Re}(3 + i)}{|3 + i|} = \frac{3}{\sqrt{10}} \quad \text{and} \quad \frac{\mathrm{Re}(4 + 2i)}{|4 + 2i|} = \frac{2}{\sqrt{5}}.$$

(iii) $f = \{([x]_R, [x^{-1}]_R) \mid x \in \mathbb{C}^*\}$ is a mapping. Indeed, if $x = a + ib \in \mathbb{C}^*$ we have $x^{-1} = (a - ib)/|x|^2$ and so

$$\frac{\mathrm{Re}\, x^{-1}}{|x^{-1}|} = \frac{a}{|x|^2} \cdot \frac{1}{|x|^{-1}} = \frac{a}{|x|} = \frac{\mathrm{Re}\, x}{|x|},$$

hence $f$ is the identity mapping.

3.46  (a) The mapping $f$ is not injective: for example, $f(1) = f(10) = 1$. It is surjective, however, since $f(0) = 0$ and for every $n \geqslant 1$ we have $n = f(m)$ where $m = 11 \ldots 1$ ($n$ terms).

(b) $n \in [1]_{R_1}$ if and only if $f(n) = f(1) = 1$, which is the case if and only if $n = 10^i$ for some $i \in \mathbb{N}$.

If $x \in [1]_{R_i}$ then $f^i(x) = 1$ so if $j > i$ we have

$$f^j(x) = f^{j-i}[f^i(x)] = f^{j-i}(1) = 1$$

and hence $x \in [1]_{R_j}$.

If $x \in [n]_{R_i}$ then $f^i(x) = f^i(n)$ so

$$f^j(x) = f^{j-i}[f^i(x)] = f^{j-i}[f^i(n)] = f^j(n)$$

whence $x \in [n]_{R_j}$. Thus $[n]_{R_i} \subseteq [n]_{R_j}$ whenever $i \leqslant j$.

$[1]_{R_1} \neq [1]_{R_2}$ since, for example, $f(55) = 10$ so $f^2(55) = f(10) = 1$ whence $55 \in [1]_{R_2}$; but clearly $55 \notin [1]_{R_1}$.

(c) Since $10^n \equiv 1 \pmod 9$ for every $n$, it follows that if

$$m = 10^n x_n + 10^{n-1} x_{n-1} + \cdots + 10 x_1 + x_0,$$

where each $x_i \in \{0, 1, \ldots, 9\}$, then $m$ is divisible by 9 if and only if $f(m) = x_n + x_{n-1} + \cdots + x_0$ is divisible by 9. Writing $f(m)$ in terms of the base 10 and applying this observation again, we see that $m$ is divisible by 9 if and only if $f^2(m)$ is divisible by 9. Applying this argument repeatedly we eventually reach some $f^i(m) = 9 = f^i(9)$. Thus $m$ is divisible by 9 if and only if $m \in [9]_{R_i}$ for some $i$.

A similar argument holds for divisibility by 3. In this case, however, at the final stage of the argument we reach some $f^i(m) \in \{1, 2, \ldots, 9\}$ which is divisible by 3. The only possibilities for $f^i(m)$ are 3, 6, 9 and so we conclude that $m$ is divisible by 3 if and only if $m \in [3]_{R_i} \cup [6]_{R_i} \cup [9]_{R_i}$ for some $i$.

(d) The reflexivity and symmetry of $R$ are obvious. For transitivity, note that if $xR_i y$ then $xR_j y$ for every $j \geqslant i$. Suppose then that $xRy$ and $yRz$. Then $xR_i y$ and $yR_j z$ for some $i, j$. If $t = \max \{i, j\}$ it follows that $xR_t y$ and $yR_t z$ whence $xR_t z$ and consequently $xRz$.

Since $f$ is surjective, so also is $f^i$ for every $i$. It follows that $R_i$ has infinitely many classes. However, for every $n$ we have $nR_i k$ for some $k$ with $1 \leqslant k \leqslant 9$ and some large enough $i$. Hence $nRk$ for some $k$ with $1 \leqslant k \leqslant 9$, and there are nine $R$-classes.

# Test paper 1

Time allowed: 3 hours
(Allocate 5 marks for each $A$ question; 20 marks for each $B$ question.)

**Section A**

**A1**   If $A, B, C$ are subsets of a set $E$ prove that
$$A \cap B \subseteq C \Leftrightarrow A \subseteq C \cup B'.$$

**A2**   If $A, B$ are subsets of a set $E$ simplify the expression
$$A \triangle [B \cap (A \triangle B')].$$

**A3**   Express as a union of intervals
$$\left\{ x \in \mathbb{R} \setminus \{1\} \, \middle| \, \frac{5 + 3x}{1 - x} \leqslant 8 + 3x \right\}.$$

**A4**   Draw the graph of the relation $\rho$ defined on $\mathbb{R}$ by
$$x \rho y \Leftrightarrow (y^2 \leqslant |x| < 1 \text{ and } x^2 \leqslant |y| < 1).$$

**A5**   Determine which of the following relations, defined on $\mathbb{C}$, are equivalence relations:
$$(a) \quad xRy \Leftrightarrow x - y = \overline{y - x};$$
$$(b) \quad xSy \Leftrightarrow x - y = \overline{x - y}.$$

**A6**   A relation $\sim$ is defined on $\mathbb{C} \setminus \{0\}$ by
$$a \sim b \Leftrightarrow \frac{a}{b} \in \mathbb{R}.$$

Show that $\sim$ is an equivalence relation and describe the $\sim$-classes.

**A7**   Let $f : \mathbb{R} \to \mathbb{R}$ be given by

90

$$f(x) = \begin{cases} \dfrac{x+1}{x+2} & \text{if } x \neq -2; \\ 1 & \text{if } x = -2. \end{cases}$$

Prove that $f$ is a bijection and find its inverse.

**A8** Given mappings $f : A \to B$ and $g : B \to C$ prove that

(a) if $g \circ f$ is injective then $f$ is injective;

(b) if $g \circ f$ is surjective then $g$ is surjective.

## Section B

**B1** In a group of 75 students, each of whom studied at least one of the subjects mathematics, physics, chemistry, it is known that 40 studied mathematics, 60 studied physics and 25 studied chemistry. Only 5 studied all three. Show that

(a) at least 25 studied mathematics and physics;

(b) at least 10 studied physics and chemistry;

(c) at most 20 studied mathematics and chemistry.

**B2** Let $\text{IN}^* = \{1, 2, 3, \ldots\}$. Define a relation $\sim$ on $\text{IN}^*$ by

$$a \sim b \Leftrightarrow (\exists m \in \mathbb{Z})a = 2^m b.$$

Show that $\sim$ is an equivalence relation and that in each $\sim$-class there is precisely one odd integer. If $a$, $b \in \text{IN}^*$ are in the same $\sim$-class, prove that either $a$ divides $b$ or $b$ divides $a$.

Let $a_0, a_1, \ldots, a_n$ be $n + 1$ distinct positive integers each less than or equal to $2n$. Prove that, for some $i, j$ with $i \neq j$, $a_i$ divides $a_j$.

**B3** Let $a, b, c, d \in \text{IR}$ with $c \neq 0$. For $x \neq -d/c$ establish the identity

$$\frac{ax + b}{cx + d} = \frac{a}{c} - \frac{ad - bc}{c^2} \cdot \frac{1}{x + d/c}.$$

Deduce that if $X = \text{IR} \setminus \{-d/c\}$ and if $f : X \to \text{IR}$ is given by

$$f(x) = \frac{ax + b}{cx + d}$$

then $f$ is either a constant mapping or is injective. What is $\text{Im } f$ in each case?

# Test paper 2

Time allowed: 3 hours

(Allocate 5 marks for each $A$ question; 20 marks for each $B$ question.)

### Section A

**A1**  Let $A = \{(x, y) \in \text{IR} \times \text{IR} \mid x^2 + y^2 - 6x + 4y + 14 = 0\}$ and let $B = \{(x, y) \in \text{IR} \times \text{IR} \mid x^2 + xy + y^2 + 1 = 0\}$. Prove that $A = B$.

**A2**  For $n \in \text{IN}$ define $n\mathbb{Z} = \{nx \mid x \in \mathbb{Z}\}$. Prove that
$$4\mathbb{Z} \triangle 6\mathbb{Z} = [4]_{12} \cup [6]_{12} \cup [8]_{12}.$$

**A3**  Express as a union of intervals
$$\left\{ x \in \text{IR} \setminus \{-1, 2\} \,\middle|\, \frac{1}{x^2 - x - 2} > \frac{1}{4} \right\}.$$

**A4**  Let $R$ be the relation defined on IR by
$$xRy \Leftrightarrow (y \geqslant [\![ |x| ]\!] \text{ and } x \geqslant |y|).$$
Sketch the graph of $R$.

**A5**  Draw the Hasse diagram of the order relation $\leqslant$ defined on the set of positive divisors of 300 by
$$a \leqslant b \Leftrightarrow a \text{ divides } b.$$

**A6**  Prove that the relation $\equiv$ defined on IR by
$$x \equiv y \Leftrightarrow x - y \in \mathbb{Z}$$
is an equivalence relation. Using the decimal representation of real numbers, show that for every $x \in \text{IR}$ the $\equiv$-class $[x]$ contains one and only one real number in the interval $[0, 1[$.

# Test paper 2

**A7** Let $f : X \to Y$ be a mapping. For $C \subseteq Y$ define the subset $C_f$ of $X$ by $x \in C_f \Leftrightarrow f(x) \in C$. Prove that if $A, B \subseteq Y$ then

$$(A \cap B)_f = A_f \cap B_f.$$

**A8** The map $f : \mathrm{IR} \to \mathrm{IR}$ is given by $f(x) = ax + b$ for constant real numbers $a, b$. For what values of $a$ and $b$ is $f$ a bijection with $f \circ f = \mathrm{id}_{\mathrm{IR}}$?

## Section B

**B1** Show that the relation $\rho$ defined on $\mathrm{IR} \times \mathrm{IR}$ by

$$(x_1, y_1)\rho(x_2, y_2) \Leftrightarrow x_1(x_2^2 + y_2^2 + 3) = x_2(x_1^2 + y_1^2 + 3)$$

is an equivalence relation.

Show that distinct points $(x, 0)$ and $(z, 0)$ belong to the same $\rho$-class if and only if $xz = 3$. Find the $\rho$-class of $(0, b)$. Describe geometrically the $\rho$-class of $(a, b)$ when $a \neq 0$.

**B2** (a) Show that

$$x^2 + xy + y^2 + 6x + 6y + 14 = (x + \tfrac{1}{2}y + a)^2 + \tfrac{3}{4}(y + b)^2 + c$$

for some $a, b, c \in \mathrm{IR}$. Deduce that the mapping $f : \mathrm{IR} \to \mathrm{IR}$ defined by $f(x) = x^3 + 6x^2 + 14x + 3$ is injective.

(b) Show that the mapping $g : \mathrm{IR} \setminus \{-1, 1\} \to \mathrm{IR}$ given by

$$g(x) = \frac{x^2 + 3x + 1}{x^2 - 1}$$

is surjective. Find $\{x \mid g(x) = 1\}$.

**B3** Let $s : \mathrm{IR} \to \mathrm{IR}$ and $t : \mathrm{IR} \to \mathrm{IR}$ be the mappings given by $s(x) = x^2$ and $t(x) = x(x - 4)$. Compute $s \circ t$ and $t \circ s$. Sketch the graphs of these composites and show that neither is an injection or a surjection.

Let $f = t \circ s$. Find the smallest $k \in \mathrm{IR}$ such that $\vartheta : [k, \infty[ \to \mathrm{IR}$ given by $\vartheta(x) = f(x)$ for $x \in [k, \infty[$ is an injection. Find a mapping $\varphi : \mathrm{Im}\, \vartheta \to [k, \infty[$ such that $\vartheta \circ \varphi = \mathrm{id}_{\mathrm{Im}\, \vartheta}$.

# Test paper 3

Time allowed: 3 hours
(Allocate 5 marks for each $A$ question; 20 marks for each $B$ question.)

## Section A

**A1**  If $A, B, C$ are subsets of a set $E$ prove that
$$(A \triangle B) \setminus C = (A \setminus C) \triangle (B \setminus C).$$

**A2**  If $A, B, C$ are subsets of a set $E$ such that $A \cup B = A \cup C$ and $A \cap B = A \cap C$, prove that $B = C$.

**A3**  Express as a union of intervals
$$\left\{ x \in \text{IR} \,\middle|\, \frac{3}{x+4} > |x| \right\}.$$

**A4**  Give two examples of relations that are both reflexive and symmetric but not transitive.

**A5**  Let $R$ be the relation defined on IR by
$$xRy \Leftrightarrow x|x| \leqslant y \leqslant |x|^2.$$
Sketch the graph of $R$.

**A6**  For every $r \in \mathbb{Q}$ with $0 \leqslant r < 1$ let $A_r = \{x \in \mathbb{Q} \mid x - [\![x]\!] = r\}$. Prove that $\{A_r \mid 0 \leqslant r < 1\}$ is a partition of $\mathbb{Q}$.

**A7**  If $f : \text{IR} \setminus \{0, 1\} \to \text{IR} \setminus \{0, 1\}$ is given by $f(x) = 1 - 1/x$ what is $f \circ f \circ f$?

**A8**  Prove that $f : \text{IN} \times \text{IN} \to \text{IN}$ given by

94

*Test paper 3*

$$f(m, n) = \begin{cases} 0 & \text{if} \quad m = n = 0; \\ 2^{m-1}(2n-1) & \text{otherwise,} \end{cases}$$

is a bijection.

## Section B

**B1** Given $f : A \to B$, let $R_f$ be the relation defined on $A$ by

$$xR_f y \Leftrightarrow f(x) = f(y).$$

Prove that $R_f$ is an equivalence relation.

Let $\mathbb{R}^* = \mathbb{R} \setminus \{0\}$ and define $f : \mathbb{R}^* \times \mathbb{R}^* \to \mathbb{R}^* \times \mathbb{R}^*$ by

$$f(x, y) = \left( \frac{y}{x}, \frac{y^2}{x^2} \right).$$

What is the $R$-class of $(-1, 1)$? Describe geometrically the $R$-class of $(a, b)$.

**B2** Determine equivalence relations on $\mathbb{R} \times \mathbb{R}$ whose equivalence classes are
   (*a*)  all lines parallel to $3x + 4y = 5$;
   (*b*)  all circles with centre $(1, 2)$;
   (*c*)  all squares with vertices on the coordinate axes.

**B3** Given mappings $X \xrightarrow{f} Y \xrightarrow{g} Z$, prove that the following statements are equivalent:
   (*a*)  there is a mapping $h : Z \to X$ such that $f \circ h \circ g = \text{id}_Y$;
   (*b*)  $f$ is surjective and $g$ is injective.
Deduce that if $\alpha : X \to Z$ is any mapping then there is a mapping $\beta : Z \to X$ such that $\alpha \circ \beta \circ \alpha = \alpha$. (*Hint:* Take $Y = \text{Im } \alpha$.)

# Test paper 4

Time allowed: 3 hours

(Allocate 5 marks for each $A$ question; 20 marks for each $B$ question.)

## Section A

**A1**  If $A, B, C$ are subsets of a set $E$ prove that

$$(A \setminus B) \setminus C \subseteq A \setminus (B \setminus C).$$

Find a necessary and sufficient condition for equality to hold.

**A2**  If $A = \{x \in \mathbb{R} \mid -2 \leqslant x \leqslant 3\}$ and $B = \{y \in \mathbb{R} \mid y^2 \leqslant y + 6\}$ prove that $A = B$.

**A3**  Express as a union of intervals the set of real numbers $k$ for which

$$\left\{ x \in \mathbb{R} \;\middle|\; \frac{(x+1)^2}{(x-1)(x-2)} = k \right\} = \emptyset.$$

**A4**  Let $R$ be the relation defined on $\mathbb{R}$ by

$$xRy \Leftrightarrow x + |x| = y + |y|.$$

Sketch the graph of $R$.

**A5**  Let $X$ be the set of mappings $f : [0, 1] \to \mathbb{R}$. Define a relation $R$ on $X$ by

$$fRg \Leftrightarrow f(\tfrac{1}{2}) = g(\tfrac{1}{2}).$$

Show that $R$ is an equivalence relation. Give two examples of elements in the $R$-class of the mapping described by $t \to t^2$.

**A6**  For every real number $r$ let $A_r = \{(x, y) \in \mathbb{R} \times \mathbb{R} \mid x - y = r\}$. Prove that $\{A_r \mid r \in \mathbb{R}\}$ is a partition of $\mathbb{R} \times \mathbb{R}$.

**A7**  Show that if $x > 1 + \sqrt{2}$ then $(x + 1)/(x - 1) < 1 + \sqrt{2}$. Show also that if

96

$A = \{x \in \mathbb{R} \mid x > 1 + \sqrt{2}\}$ and $B = \{x \in \mathbb{R} \mid 2 + 2\sqrt{2} < x\}$ then the mapping $f : A \to B$ given by $f(x) = (x^2 + 1)/(x - 1)$ is injective.

**A8**  Let $g : \mathbb{R} \to \mathbb{R}$ be defined by
$$g(x) = \begin{cases} 1 - x & \text{if} \quad x \geqslant 0; \\ (1 - x)^2 & \text{if} \quad x < 0. \end{cases}$$
Show that $g$ is a bijection. What is $g^{-1}$?

**Section B**

**B1**  If $E$ is a non-empty set then a non-empty collection $\mathscr{F}$ of subsets of $E$ is called a *filter* if
(a)  $F \neq \emptyset$ for every $F \in \mathscr{F}$;
(b)  if $F_1, F_2 \in \mathscr{F}$ then $F_1 \cap F_2 \in \mathscr{F}$;
(c)  if $G \supseteq F$ and $F \in \mathscr{F}$, then $G \in \mathscr{F}$.
Verify that the collection of all subsets of $E$ that contain a given element of $E$ is a filter.

Show that if $\mathscr{F}$ is a filter and $X$ is a subset of $E$ such that $X \cap F \neq \emptyset$ for every $F \in \mathscr{F}$, then
$$\mathscr{F}_X = \{Y \mid (\exists F \in \mathscr{F}) \, Y \supseteq X \cap F\}$$
is also a filter. Show also that $\mathscr{F} \subseteq \mathscr{F}_X$, and that the inclusion is strict whenever $X \notin \mathscr{F}$.

**B2**  Let $E$ be the set of months in the year. Show that the relation $R$ defined on $E$ by $xRy$ if and only if $x, y$ start on the same day of the week is an equivalence relation on $E$. Determine the equivalence classes for an ordinary year and for a leap year. How many Friday the thirteenths can there be in a year?

**B3**  Let $X = \{1, 2, \ldots, n\}$. Show that a mapping $f : X \to X$ is such that $f \circ f = f$ if and only if the restriction of $f$ to $\operatorname{Im} f$ is the identity map.
Suppose now that, for $1 \leqslant r \leqslant n$,
$$E_r = \{f : X \to X \mid f \circ f = f, \; |\operatorname{Im} f| = r\}.$$
Prove that
$$|E_r| = \binom{n}{r} r^{n-r}.$$